MONKEYS WILL NEVER TALK... OR WILL THEY?

Clifford Wilson

M.A., B.D., Ph.D

Senior Lecturer in Education
(Psycholinguistics)
Monash University, Victoria, Australia

MASTER BOOKS
A Division of CLP
San Diego, California

THE AUTHOR

Clifford Wilson, Ph.D.

Dr. Wilson is a council member of the Commercial Education Society of Australia and in 1971 was honored as "An Outstanding Educator of America" while teaching at the University of South Carolina. Currently, he is Senior Lecturer in Education at the Monash University in Melbourne, being lecturer-in-charge of Psycholinguistics.

While Professor of Early Childhood Education at the University of South Carolina, he compiled the Language Abilities Test extending from early childhood to adult.

Dr. Wilson is eminently qualified to write a book dealing with language abilities and has done an excellent job in the collection and presentation of studies, surveys, and scientific data concerning speech, its acquisition, and innate qualities.

Although best known for his archaeological studies and his book refuting the "gods from outer space" theory, Dr. Wilson's primary field is that of psycholinguistics.

MONKEYS WILL NEVER TALK ... OR WILL THEY?

A Survey Dealing With —

The Experiments with Chimpanzees
The Principles of Animal Communication
Apes and babes — a comparison and a contrast

DEDICATION

A wife who has seen twenty of her husband's books published and still takes an interest in a new one, deserves a gold medal.

Thank you again, Avis!

ACKNOWLEDGMENTS

Special acknowledgment is due Graham Morris, Stewart Jackel, and Margaret Liddell, Education Officers of the Melbourne Zoo Education Service in Australia. Their cooperation in discussion, research, and the provision of a stream of current articles and journals has been a tremendous help.

Dr. Paul Ackerman, of the Psychology Department at the Wichita State University in Kansas, has helped greatly by his thought-provoking suggestions as the manuscript progressed to its final form.

CONTENTS

Introduction 9

PART I: CHIMPS MAKE SIGNS
TO TELL THEIR TALE

1. Talking Monkeys? 13
2. The "Talking Chimpanzee" From Nevada 21
3. Washoe: A Contrast With Children's
   Language Learning 28
4. Sarah and Her Plastic Signs
   for "Dumb" Humans 38
5. How's Your Yerkish? 48

PART II: THE BEES DANCE
TO TELL THEIR TALE

6. What Is Communication? 59
7. The Bees Dance and the Birds Sing 67
8. Gestures of Gorillas and Sounds of Dolphins ... 75
9. Animals and Problem-Solving 83
10. Pecking Pigeons Are Not Babbling Babies 90

PART III: THE SILENT APE
AND THE BABBLING BABE
A 23-POINT ANALYSIS

11. How Chimps and Babes Communicate 99
12. What About Eyes and Ears? 116
13. Unique Information-Processing 131
14. Words Can Be Used in Different Ways 146
15. How Language Is "Brought Together" 160
Conclusion 174
Bibliography 175

INTRODUCTION

At regular intervals dramatic challenges are thrown out against basic Christian beliefs. Probably the best known was that by Charles Darwin, but there have been others.

In this volume our special interest is in speech and language, and we shall analyze some experiments which at first appear to have challenged man's uniqueness as a word-using being. The issue is very important and highlights essential differences between behaviorists and certain other schools of psychology.

It also highlights a basic difference between behaviorists and Christian creationists. Twenty years ago Professor B. F. Skinner and his "pecking pigeons" seemed to put aside the need to think of God and creation, for those pigeons apparently demonstrated learning techniques that could be transferred to other living creatures, including man. Noam Chomsky's devastating answer stemmed the tide, and eventually it was widely recognized that the "pigeon" learning principles did not apply to human language acquisition and learning after all.

The behaviorists were not to be silenced, and a series of different experiments with chimpanzees heralded a new attack against those who held to man's uniqueness as a word-using being. The evidence is sufficient to demand serious attention, if only because it has led to exaggerated claims by some investigators.

We shall see that there are basic differences of principle in the methods utilized by chimpanzees when compared with those used by human infants. Ultimately, the real difference is that humans have inherent language/speech abilities, and chimpanzees are not so endowed. We support this conclusion by evidence to counter that put forward by the researchers with chimpanzees. This evidence is highly important in the fields of psychology, psycholinguistics, and education.

PART I
CHIMPS MAKE SIGNS TO TELL THEIR TALE

**A Survey Of Five Major Experiments
With Chimpanzees**

Chapter 1
TALKING MONKEYS?

History has many examples of humans trying to teach monkeys to speak, and popular writers often tell us we are at the point of a breakthrough. Today the case seems stronger than ever, with a series of experiments involving chimpanzees. They have demonstrated communication abilities of a type unknown 50 years ago.

The road has been long and arduous, but researchers have persisted, learning from failures and patiently applying new techniques in their endeavors to probe the minds of chimpanzees in captivity. Their patience has been rewarded in that they have demonstrated unexpected intelligence in their subjects, but their results are not as clear-cut as some popular writers have claimed.

This is especially true in the realm of speech as it is practiced by humans. To assess those results we need to know something of man's efforts to teach language to "lesser creatures."

SOME ATTEMPTS TO TEACH LANGUAGE TO APES

In 1933 there was a report that Dr. Winthrop Kellogg and Mrs. Luella Kellogg had brought a female chimpanzee named Gua into their home, with the hope that she would learn English at the same time and in the same way that their infant son acquired that language. Over a period of some 16 months they found that Gua was able to understand about 100 words, but at no time did she attempt to speak any of those words.

With humans there is both receptive and expressive language, and each of these can be broken up into several subdivisions. It is the total of those subdivisions that make up pure language as used by humans. There is no doubt that some animals can utilize some of those features, but only humans utilize them all: that is the distinction between "speaking" humans and non-speaking animals.

This is relevant when we consider that the chimpanzee Gua could understand a range of English words. She had "receptive language" for those hundred words, but actually there is nothing very remarkable about that. I have a dog named Bunty, and when I tell Bunty to sit, she sits. When I tell her to roll over she does so, for she knows that she will be rewarded by having her tummy rubbed. In earlier years I have had other dogs which I have taken the trouble to train so that they would respond to a number of words, and of course this teaching of particular words to animals is known right around the world.

In such training both negative and positive forms of reinforcement are used by some researchers with rewards that are pleasant, or — with some experimenters — punishments that are not so pleasant. Whatever the methods, animals can be taught to respond to a large number of human words.

We do not for a moment deny that Gua's accomplishment was considerable, but this is to be expected when intensive training is given to chimpanzees, for they are recognized as being highly intelligent. Some researchers argue that the porpoise is more intelligent than the chimpanzee, but all will acknowledge that in the animal kingdom the chimpanzee is very high in mental capacity.

Not only that, but chimpanzees are renowned for their sociability and for their forming of strong attachments to the humans with whom they are associated. They are sensitive creatures and respond to human affection.

However, there is always the possibility that they will prove difficult or even dangerous, for a full grown chimpanzee is about five times as strong as the average man of the same weight. A full grown chimpanzee weighs approximately 120 pounds.

Gua demonstrated that a chimpanzee had a good capacity (by "animal standards") to understand that certain sounds (words) could be related to identifiable objects or actions. However, her attainment was greatly inferior to that of the normal human infant given similar exposure to language.

In *Readings in Animal Behavior* (1973) there is an article by Dr. Kellogg entitled "Communication and Language in the Home Raised Chimpanzee."[1] At the time of writing he was Professor in Experimental Psychology at Florida State University, Tallahassee. He discusses the way oral speech develops in the human infant and points out that the deaf mute fails to speak because he does not hear the acoustic patterns of which words are made up: "He has no sound patterns to follow, no models to imitate." As Dr. Kellogg points out, the human infant needs a normal ear, a normal brain, and speech organs, as well as the continuous hearing of spoken language, plus a great deal of imitation if the speech process is to be completed. We shall see that this is very relevant to our discussion about "talking monkeys."

The child of deaf parents will learn to speak if he himself has normal hearing and is exposed to and interchanges with language. It is not enough for him to simply hear language, as was the case with one child (reported by Dan Slobin) who spent much time during the first three years of his life in front of a television set. Though he was exposed to a great deal of human speech, it was found that he had not learned to speak. The acquisition of speech demands interaction with speech as well as actual exposure to human language.

GUA: "LITTLE EVIDENCE OF VOCAL IMITATION"

Because of the experiment with Gua, Dr. Kellogg was in a firsthand position to analyze the possibility of teaching language to a chimpanzee by giving it the same exposure as that of a human child. Kellogg himself says that at first glance the results were disappointing:

> . . . even in the experimentally controlled environment in which a home-raised chimpanzee [Gua] is given the same linguistic and social advantages as a human baby, the chimp displays little evidence of vocal imitation. Despite its generally high level of imitative behavior, it never copies or reproduces human word sounds.[2]

In that same article Kellogg tells us that Gua made no sounds "without some definite provocation. . . and in most cases the stimulus was obviously of an emotional character." He refers also to a similar experiment whereby Mrs. N. Kohts in Moscow tried to teach language to the chimpanzee Joni, and, like Gua, Joni did not give any evidence of trying to reproduce human vocalizations.

Kellogg points out that Gua had a series of behavior patterns which humans could interpret. Thus she would remove the bib from her neck to indicate that she had finished eating. She would throw herself on the floor, and humans knew she was sleepy or tired, and would go to sleep at once when put to bed. She would climb into a high chair, and this meant she was hungry. She would protrude her lips towards a cup to indicate she wanted a drink and would push the cup away when she had enough.

Although such learned "signals" pointed to Gua's ability to bend humans to her will, they were not "language" in the pure sense. They were a form of communications relating to concepts limited to the self-interest of the chimpanzee, and they highlighted the great chasm between the "language" of Gua and that of a young child.

THE CHIMPANZEE THAT "SPOKE" FOUR WORDS

Another experiment dated to the early 1950's. Dr. Keith Hayes and Mrs. Cathy Hayes raised the female chimpanzee, Viki, in their home, and in three years of training she learned four sounds that approximated English words.[3] Over a total period of six years, Viki used those four words and several other sounds that did not resemble English words, but were used for specific requests.

This experiment highlighted the fact that there are quite serious physiological barriers against chimpanzees vocalizing in the way that is characteristic of human speech.[4]

Viki behaved in the home of Dr. and Mrs. Hayes as a highly intelligent animal, exploring her world and using various household appliances and tools. In some ways she was more advanced than the 18-month boy to whom she was introduced when she was 9 months old, though there was one important difference. The boy could speak a dozen words but Viki could not speak at all. In her earliest weeks she had made sounds not unlike those of a human baby, rather like "ooo-ooo-ooo" and "uh-uh-uh," but she did not develop this babbling type signal as she grew older.

This was a significant difference from the progress of the young child. Viki showed little interest in sound play, and this is in direct contrast to the human infant, for he practices and experiments with sounds constantly as he goes through regular stages of vocalization and acquisition of language. The chimpanzee has hearing as keen as that of man, and his vocal mechanism is in some ways similar, but when it comes to the fineness required for the making of human speech sounds there are dramatic differences, and the available data indicate that "the nonhuman primates would not be capable of producing human speech even if they had the requisite mental ability."[5]

The Hayes were able to teach Viki to say "Mama," but only after many hours of instruction. They had no response whatever to merely repeating the word over and over again to the little chimp, for she did not imitate the word as they hoped. They persevered by holding her mouth and covering her lips in the proper position so that the sound would be produced, and eventually Viki could mouth the word without help.

However, even when Viki had "learned" this word she would only mouth it while holding her lips with her hand, in the same way the Hayes had done. In a similar way she learned two other words, "cup" and "Papa," and used both with proper meaning. If she wanted a drink she would actually say "cup." For each of these words she utilized the same vowel, in a hoarse whisper. Possibly she also learned the word "up." However, there was a tremendous amount of time and effort involved in teaching Viki those few words, and clearly the prospect of a chimpanzee mastering the dictionary requirements for spoken human language is, to say the least, remote. Viki's achievement was simply not in the same class as that of a normal child of the same age who by that time would know several hundred words.

NO REARRANGED SEQUENCES OF WORDS

In that same time Viki had learned to respond to 50 or more commands, announcements, and questions. However, her obedience was not only to the actual spoken content of a particular commandment, but it also partly resulted from the tone of voice or a gesture of the Hayes. Nor did she understand words rearranged in new sequences, and she became confused if sentences were too long. To her, words were signals rather than symbols of language in the sense that human beings understand them. She had the capacity to understand those symbols in isolation, when they were related to immediate situations, but she did not display a human-type understanding when it came to abstract expressions.

We have seen that in the various experiments with apes, they have been found to be social creatures with relatively high intelligence when measured against other animals. Some have even made some limited use of abstract symbols that make up man's language, "but in comparison with men their talent in this direction is extremely limited."[6]

In addition to his experiment with Gua, Dr. Kellogg also discusses the experiments of some other researchers who had raised chimpanzees, and he refers to the consistent failure of apes to utilize vocal chords for human speech. He recognizes that the best of the "early" attempts was that of Dr. and Mrs. Hayes with Viki. Kellogg suggests that Viki's use of the four words "Mama," "Papa," "cup," and possibly "up," represented the acme of a chimpanzee's achievement in its capacity to produce sounds of human speech, and he reminds us that "even these were learned only with the greatest difficulty."

Dr. Kellogg further points out that "even after she could reproduce them, the animal's words were sometimes confused and were used incorrectly."

He also suggests that possibly the most important aspect of the Hayes' work was not that their chimpanzee could produce a few human sounds, but that they showed that such sound patterns were extremely hard for an ape to master. Viki did not master them easily or naturally, and even when she had learned them she had trouble in keeping the patterns of the speech sounds in proper order.

"ALL HAVE FAILED DISMALLY"

Dr. Kellogg is not the only researcher who comments on Viki's capacity in language learning. Another is

Professor W. H. Thorpe, of Cambridge University in England, who concluded, "The result was that in six years Viki learned only four sounds that approximated to English words, and even the approximation was somewhat tenuous."[7] Professor Thorpe had spent some days with Dr. and Mrs. Hayes in 1952.

The Hayes had worked very hard with Viki for six long years, but her language learning was disappointing in a number of ways. David McNeill includes this experiment to teach chimpanzees to talk under the subheading of "Can Apes Talk?" and then states, "but all have failed dismally."[8] McNeill was not including the experiments conducted at a later time, especially those with the chimpanzees Washoe, Sarah, and Lena. We shall analyze these as we proceed.

Footnotes: Chapter 1

1. *Readings in Animal Behavior*, 1973, pp. 309 ff.
2. *Ibid.*, p. 311.
3. *The Ape In Our House*, C. Hayes.
4. The classical article in this connection is "Primate Vocalizations and Human Linguistic Ability" by Philip Lieberman, reprinted from *The Journal of the Acoustical Society of America*, Vol. 44, No. 61, December, 1968, pp. 1574-1584.
5. *Ibid.*, p. 1580.
6. *How Animals Communicate*, B. Gilbert, p. 133.
7. In *Non-Verbal Communication*, Chapter entitled "The Comparison of Vocal Communication in Animals and Man," p. 56.
8. In *The Acquisition of Language*, p. 54.

Chapter 2
THE "TALKING CHIMPANZEE" FROM NEVADA

The husband and wife team of Dr. R. A. Gardner and Dr. B. T. Gardner began their famous experiment in June, 1966, in association with the University of Nevada. They knew of the relative failure of earlier experimenters and learned from them. They decided to use a gestural language common to many deaf people, and so utilized the American Sign Language as it is used by the deaf in North America. They were aware that many problems faced them, including the fact that "the vocal behavior of the chimpanzee is very different from that of man." They knew that chimpanzees make a number of different sounds, but that generally they occur in situations involving high excitement, and the sounds themselves tend to be specific to those exciting situations.[1] If chimpanzees are undisturbed they are usually silent, and the Gardners knew that a chimpanzee would be unlikely to make refined use of its vocal chords.

They had studied the intensive work undertaken by Dr. & Mrs. Hayes with the chimpanzee Viki, and they had concluded that the Hayes' method was not appropriate with such a non-human subject.

USING HANDS FOR "TALKING"

The Gardners realized that the hands were prominent in the behavior of chimpanzees, and that they were well-known for their ability to cope with manipulatory mechanical problems. In their article they point out that chimpanzees that have had extensive contact with human beings have displayed a wide variety of communicative gestures.[2] They were influenced by this in their selection of the American Sign Language as the chosen medium of communication. They reasoned that this would probably be appropriate because of the anatomical similarity between the hands of the chimpanzee and of man. Chimpanzees can use human tools and devices with considerable skill, for their motor capacities are relatively well developed. However, as they are unable to vocalize with the precision required of human speech, it seemed that the hands would be appropriate as the medium through which "language" could be communicated.

This was a logical choice and fits in with a well-established practice where certain types of animals are chosen by psychologists according to the suitability of those "subjects" for particular experiments. Rats are very capable of pressing levers, and so they are highly suitable for the lever-type operations involved in some of Professor B. F. Skinner's experiments at Harvard University. Dogs were ideal for Pavlov's experiments with bells, food, and saliva flow. So, as the Drs. Gardner say, "We chose a language based on gestures because we reasoned that gestures for the chimpanzee should be analogous to bar-pressing for rats, key-pecking for pigeons, and babbling for humans."[3]

They "accommodated" the chimpanzees even further. There are two different systems of signs used by deaf people: one utilizes gestures, and the other is finger spelling whereby words are actually spelled out by the letters of the alphabet. Obviously a person cannot use finger spelling unless he can read in the human sense,

and this involves knowing an alphabet and the great
varieties of possible associations as consonants and
vowels are brought together in the complexities of written words. The problems of teaching an alphabet and
then words to a chimpanzee are so great that they hardly
need be commented on.

WASHOE AND THE AMERICAN SIGN LANGUAGE

So it was that the Gardners chose the American Sign
Language (A.S.L.), which has some similarities to picto-
graphic writing. One other advantage of this system was
that it is widely used in North America, and this meant
that users of this sign language could take part in the
training of Washoe.

Washoe is actually named after Washoe County, the
County where the University of Nevada is situated. The
Gardners estimated from her weight and dentition that
she was between eight and 14 months of age when she
came to them in June, 1966. It is relevant to notice that
this was the earliest age that Washoe could reasonably
be expected to learn "language" by the system adopted
by the Gardners: "She did not have her first canines or
molars, her hand-eye coordination was rudimentary,
she'd only begun to crawl about, and she slept a great
deal."[4]

The Gardners themselves acknowledged that they
could accomplish little with her during the first few
months, for, as the above quotation makes clear,
Washoe's hand-eye coordination was rudimentary. As
we shall see, the "language" that Washoe was learning
depended essentially on hand-eye coordination. Visual
stimulus was fundamental, and her coordination with
motor activities was part of the language teaching process.

First, successful attempts were made to win Washoe's
friendship. Her confinement was minimal, being about
the same as it would be for human infants. The humans

brought into contact with her were her friends and "playmates," as well as being those who provided for her. They entered into a large number of games and activities to ensure maximum interaction with Washoe.

At least one person was with Washoe during all her waking hours, and she lapped up human affection. They all utilized the American Sign Language extensively in her presence, with many interesting activities, just as though they were chattering at a human infant all day. Spoken English was rejected, to ensure that the chimpanzee would pay sufficient attention to their gestures.

THE IMPORTANCE OF IMITATION AND IMITATING

Washoe's environment was not a silent one however, for there were various vocalizations such as laughing and emitting sounds of pleasure or displeasure. Whistles and drums were used in various imitation games, and clapping hands for attention was common. The Gardners tell us, "The rule is that all meaningful sounds, whether vocalized or not, must be sounds that a chimpanzee can imitate."[5] This leads to the important fact that the most signifcant aspect of Washoe's training was in the realm of imitation.

In Chapter 11, dealing with the stages of language acquisition by young children, we make the point that imitation is not the essential factor in the acquisition of language by young children. No one ever taught a child to say, "I falled down and I breaked my arm," or, "I saw three mouses." These are simply the child's making of new words by following the principle of regularizing. Later he will learn that those particular words "falled," "breaked," and "mouses" give place to "fell," "broke," and "mice," simply because irregular words take the place of the words that would be expected if the principle of regularizing was followed in those cases. This is one of a number of indications that imitation is not the primary role in the acquisition of language by young children.

"VISUALLY GUIDED IMITATION"

This is not so with apes. As the Drs. Gardner state, "The imitativeness of apes is proverbial, and rightly so. Those who have worked closely with chimpanzees have frequently remarked on their readiness to engage in visually guided imitation."[6]

Note especially those words, "visually guided imitation." Human language depends largely on auditory factors rather than visual. We are referring to primary human language rather than to secondary language, but again this is highly significant as a point of contrast with the acquisition and learning of language by chimpanzees. This is an essential point of difference in what is actually being acquired by chimpanzees when compared with the acquisition of language by young children.

In an extensive quote by Dr. Gardner, this statement as to the imitative tendency of chimpanzees is relevant: "It seems to be controlled chiefly by visual stimuli. Things which are seen tend to be imitated or reproduced." Dr. Gardner goes on to make this significant statement, "What is heard is not reproduced. Obviously an animal which lacks a tendency to reinstate auditory stimuli — in other words to imitate sounds — cannot reasonably be expected to talk."[7]

Here we have an essential difference from human language. The human child has that capacity to imitate sounds, but the chimpanzee's capacity is greatly inferior. Actually the parrot has it to a much greater degree than the chimpanzee, but neither creature has the cognitive processes which the human child has. Those greater cognitive abilities, together with the ability to reinstate auditory stimuli after the sound itself has ended, give the child a potential for language beyond that of any chimpanzee.

The Gardners knew that in their work with Viki the Hayes had devised a game whereby Viki would imitate certain actions on the command "Do this." They would use these words when they wanted Viki to undertake certain actions that were visually guided. The Gardners recognized that this could be an effective means of teaching Washoe the American Sign Language, and so they set out to teach her to follow the same pattern.

IMITATION BROUGHT REWARDS

To get Washoe to imitate was not difficult at all, for she did that almost spontaneously—as any adult can ascertain by a visit to a zoo. However, it was a different matter to get her to imitate on the command "Do this," and it was not until 16 months had passed that they had any degree of success. Their eventual success was by using the reward of tickling, which Washoe especially enjoyed.

The experiments were somewhat disappointing as methods of introducing new signs into Washoe's vocabulary. Nevertheless, in the measure of success they did have, imitation as such continued to be important, and they say, "As a method of prompting, we have been able to use imitation extensively to increase the frequency and refine the form of signs."[8] In the earlier stages they expected Washoe to make a sign roughly equivalent to the "correct" sign. By the imitative process they could take her hands and form them into the desired configuration, and this method has been frequently followed. As time went by, Washoe's manual "diction" improved, and prompting became less necessary.

The words that Washoe accumulated were associated with her personal desires, and basically there was a reward-gaining factor involved in this method of pleasing her trainers. She was prepared to learn and to use the signs correctly in order to gain that reward.

Over a period the signs could be transferred to other members of the class of referents. Thus Washoe has been able to transfer the sign for "dog" to the sound of barking by an unknown dog. This points to an intelligence not unlike what the Swiss psychologist Jean Piaget has termed the sensorimotor stage of intelligence in the young child, though it is by no means as elaborate.

Washoe also learned to understand relatively simple combinations such as "listen eat", which the Gardners would use when an alarm clock was sounded, thereby signaling meal time. She would signal "listen dog" when she heard the barking of an unseen dog, again showing a high degree of animal intelligence, including the capacity to associate one concept with another. Such intelligence has not been denied by researchers, for the maze learning of rats, the pecking of pigeons, and so much more, all point to the highly complex nature of intelligent operations that can be undertaken by non-human beings.

Even at this level of "intelligence" there is a great contrast between animals and children, especially after the child has reached two years of age. The difference is also dramatic in the realm of language, as we shall see in our next chapter.

Footnotes: Chapter 2

1. From article "Teaching Sign Language To A Chimpanzee," by the Drs. R. A. & T. Gardner in *Language in Thinking* (Reprinted from *Science*, vol. 165, No. 3894, 1969).
2. *Ibid.*, p. 19.
3. *Ibid.*, p. 20.
4. *Ibid.*, p. 22.
5. *Ibid.*, p. 24.
6. *Ibid.*, p. 24.
7. *Ibid.*, p. 24.
8. *Ibid.*, p. 25.

Chapter 3
WASHOE: A CONTRAST WITH CHILDREN'S LANGUAGE LEARNING

The Gardners point out that Washoe's earliest signs used were simple demands, but that most of the later signs were names for objects. This is somewhat different from the pattern shown by young children in the acquisition of language. At 16 months of age approximately 50 per cent of the child's words are one-word sentences, often demanding action but expressed by nouns. Action words usually come considerably later.

Even when the child moves on from these "one-word sentences" (known as holophrastic expressions) to two words, it is likely to be an adjective and a noun, still involving only one concept such as "good doggy," and not "listen dog" as in one of Washoe's expressions. This latter involves first the action of listening, and then the fact that what is making the noise is a dog. When she made the signs for "listen eat" she was involving two action concepts — listening that was associated with a bell and eating of the food that would follow the sound of the bell. As we have said, this is different from the developing child where the first expression will probably be in the form of a noun, and then with two words the new word is likely to be an adjective qualifying that noun. It is probable that the child is still dealing with one basic concept and not with two, as with the chimpanzee.

This sort of analysis can be applied to the list of signs that the Gardners give. In their initial report they referred to 34 such signs, in the order that Washoe acquired them.[1] In the main, they involved reward-giving. This can be demonstrated by looking at the very first four signs: "come gimme," "more," "up," and "sweet." Other examples given by the Gardners include "come tickle" and "gimme sweet." The sign for "more" could be asking for the continuation of activities such as swinging or tickling, second helpings of food, etc. The sign for "up" was used when Washoe wanted to be lifted up to reach objects such as grapes; and "sweet" was when Washoe wanted dessert or candy. This sort of reward-gaining behavior is dramatically different from the language acquisition pattern of a young child.

In addition to "come," "gimme," "more," "up," and "sweet," other signs in order of acquisition were "open," "tickle," "go," "out," "hurry," "here-listen," "toothbrush," "drink," "hurt," "sorry," "funny," "please," "food-eat," "flower-blanket," "dog," "you," "napkin-bib," "in," "brush," "hat," "I-me," "shoes," "smell," "pants," "clothes," "cat," "key," "baby," and "clean."

Despite this ability to use a number of signs correctly, not all linguists are convinced that Washoe has in fact learned to use genuine words. However, most who have seen the film that the Gardners have made of Washoe accept that she does use actual "words" by her signs.

ANIMALS "LEARN" WHEREAS CHILDREN "ACQUIRE" LANGUAGE

Nevertheless, this does not mean that Washoe acquires speech along the lines of human acquisition. David McNeill points out that with young children holophrastic speech consists mainly of three kinds of utterance, these being (1) the inventions of the child, (2) nouns, and (3) adjectives.[2] Support words, such as the

preposition "in" that Washoe used, are not found in the
vocabulary of the young child, and in the so-called
"words" of Washoe there were no inventions of her own,
as with the young child. We have already seen that with
Washoe action words predominated rather than nouns
and adjectives. As David McNeill points out, among
Washoe's first 34 words eight were verbs and two were
prepositions, as against 16 that were nouns, three adjectives, and no invented words.

Verbs and prepositions are the last to appear in
human speech, but they were the first with Washoe.
Children are exposed to verbs and prepositions all the
time and yet they use them late, whereas Washoe (and
the chimpanzee Sarah also—see Chapter 4) used them
early, simply because they were "learning" and not "acquiring" according to an inherent capacity. They were
participating in a communication exercise that involved
a reward-gaining performance, whereas children simply
follow a pattern of language acquisition that appears to
have been there from man's earliest time.

The Gardners reported that Washoe did not combine
signs in a definite order, and that basically her words
were produced in their order of importance to her. This
also is explained by the fact that it was a reward-gaining
experience. There was no true grammatical relationship
as such, and this is the opposite of the experience of
young children as they acquire and utilize the grammar
of English or any other language.

The Gardners' experiment produced other interesting
and relevant data. They point out that in the first four
months of the study the chimpanzee was slightly
ahead of their child in the number of phrases to which it
could correctly respond. They suggest this was due to
her superior locomotive ability, for obviously the child at
first could not comply with such directions as, "Get up
on the chair." However, during the last five months of
the experiment the child surpassed the ape in comprehension. At the end of 9 months the child could respond
to 68 specific responses and the chimpanzee to 58. The

chimpanzee was overtaken by the child, and the child's acceleration was at a much faster rate than that of the chimpanzee.

THE ROLE OF BABBLING WITH CHIMPANZEES

There were other differences which the experiment highlighted, such as the role of babbling. Babbling is one of the early forms of pre-language expression with the human infant, but this decreases as other stages of language acquisition are reached. The opposite was true with Washoe. They found that at first there was no "manual babbling" by random use of the sign language by which she was surrounded. There was no great activity in the early months, but manual babbling actually increased as the months went by. However, it eventually declined and, according to the Gardners, "Its function in the total process of acquiring A.S.L. seems to have been comparatively slight and rather doubtful."[3]

Washoe's pattern was obviously quite different from the processes involved in children's babbling. After she had learned certain signs, if she could not get what she wanted, she would sometimes "burst into a flurry of random flourishes and arm waving."[4]

The Gardners themselves encouraged Washoe's manual babblings by their own responses, such as clapping and smiling, and letting the chimpanzee know they were pleased. If at times the "babbled gesture" bore some similarity to a sign in the American Sign Language, they would make the correct form of the sign, then use it in some appropriate activity. The sign for "funny" was probably acquired in that way, and by various aspects of imitation they would agree with Washoe that something was "funny." They state that "eventually Washoe came to use the 'funny' sign spontaneously in roughly appropriate situations."[5]

However, a child does not come to use a word in only

"roughly appropriate situations," but is soon using it with great specificity.

In a number of other ways the methods used for teaching language to Washoe were quite different from those used with children. The Gardners talk about instrumental conditioning and state frankly that one of their objectives was to teach Washoe as many signs as possible by whatever procedures they could enlist. They did not hesitate to use "conventional procedures of instrumental conditioning."[6] They made much of tickling as a reward, and this for instance was an easy way to teach her the sign for "more:" they would stop tickling her until she responded with the appropriate sign.

As time went by they expected a closer approximation to the correct sign than the crude attempt they at first accepted, and they introduced imitative prompting. That "more" sign was soon being used satisfactorily, though at first "it was quite specific to the tickling situation." Later they used this to apply to other activities as well.

This highlights a basic difference from the acquisition of language by young children, as discussed above: children do not acquire or utilize language simply because of possible rewards. They acquire language because they are so designed as to be able to grasp its principles from their surrounding environment, and then they utilize it in everyday activities. Instead of it being a great effort to teach children language, it would in fact be difficult to teach them not to use language. This is a great contrast: whereas it is extremely difficult to teach any form of language to a chimpanzee, it would be even more difficult to restrain children with normal faculties from speaking!

Signs used in experiments such as those of the Gardners have been called "cheremes," as distinct from words that can be analyzed into phonemes or sound units. There are 55 cheremes in the American Sign Language (A.S.L.). In an article in *Non-Verbal Communication* Professor W. H. Thorpe of Cambridge points out that nineteen of these identify the configuration of the hand or hands making the sign, while 12 of the signs identify the place where the sign is made, and the other 24 relate to the hand or hands. They are of course manual, as contrasted with the vocal output in human speech.

CATEGORIZING WASHOE'S SIGNS

Washoe has demonstrated a considerable degree of efficiency, and, when she made errors, in a majority of cases the errors were within a particular grouping. Thus her items could be classified into categories: her vocabulary included "brush," "clean," "soap," "comb," "oil," "lotion," and "toothbrush," all of which were associated with grooming. "Bird," "bug," "cat," "cow," and "dog" could be categorized as members of the animal family, while "banana," "drink," "fruit," "meat," and "sweet" were all associated with food. Professor Thorpe points out that in one set of experiments in which Washoe made errors in 12 trials, seven out of the 12 times she gave signs for other items that were in the relevant category. In that particular incident she was dealing with articles for grooming, but similar results were obtained for other categories as well.

In the three years following the commencement of the experiment, Washoe mastered approximately 87 signs similar to those which make up the A.S.L. We have seen that in a comparable period, with extensive training under Dr. & Mrs. Hayes, Viki acquired the ability to use only four vocalizations, and even one of those was doubtful. This demonstrates that signs of the A.S.L., and similar cheremes, have made possible a breakthrough in the acquisition of a form of communication suitable for chimpanzees, when reward-reinforcement is sufficiently strong.

The achievements by Washoe go beyond the actual acquisition of 87 different A.S.L. signs, for we have seen that she was able also to combine signs so as to give a two-word message. In just over three years—38 months from April, 1967—294 different two-sign combinations were used by Washoe. The vocabulary of these two-sign combinations was actually divided into two groups. Because of the terminology often used with the acquisition of language by young children, one group was called "pivots," these being words used in the bulk of the two-word combinations. (The term "pivot" has now fallen out of favor with psycholinguists.)[7] The larger group of lesser-used words consisted of those words which were almost always paired with one of the "pivots."

It was then further found that 12 signs were the most frequent, and in order of frequency these were the signs for "come-gimme," "please," "you," "go," "hurry," "more," "up," "open," "out," "in," and "food."

In their 1971[8] report, the Gardners put Washoe's 87 signs into six categories as follows:

Category	Typical Signs
Appeal	— gimme, hurry, more
Location	— in, out, up
Action	— come, hug, tickle
Object	— banana, clothes, pencil
Agent	— me, you, Washoe
Attribute	— black, enough, sorry

These overlapped to a considerable extent. It is possible that the first of them (appeal) can be fitted into the other five categories, these five being the categories suggested by Roger Brown in *The Acquisition of Language by Children* (1970).

Washoe could also elaborate her desires to a limited extent. If she showed a sign such as "please" or "come," and was then asked a question by her human partner, "What you want?", she could add a second word which would be the correct answer. It might be, "Please out," or "Come open," or "More tickle," and so she was answering questions according to the limitations imposed by the signs she had already learned.

WASHOE'S SERIOUS LIMITATIONS

The Gardners frankly state that Washoe was by no means expert with the signs she "learned." With new signs, and often with old ones as well, she would lapse into what they called "poor diction." Also, if she did not want to be corrected, she would at times respond in various ways such as throwing a tantrum, running away, or even biting her tutor.

In contrast, the normal young child does not make mistakes with learned words to the extent that Washoe did, nor do most children go into tantrums or bite their tutors when correction is offered! Nor of course do normal young children need signs to express themselves. They simply open their mouths and create very many new utterances and combinations to a degree unthought of for Washoe. She was limited to the relatively small number of manual-visible signs she learned by the diligent application and great patience on the part of her instructors, over a considerable period of time.

The most that can be said is that Washoe has learned a limited communication system that has analogies to the American Sign Language. When compared with young children, and even with deaf children, the achievements of Washoe leave no doubt that humans have a capacity for acquiring and learning language that is dramatically greater than that of the chimpanzee.

Washoe was eventually moved to the University of Oklahoma and her "education" continued, together with other chimpanzees taught by the same methods.[9] Possibly they will also be capable of learning from each other, just as other monkeys have learned the art of swimming, and even a crude form of fishing, from their peers. Such accomplishments are of great interest, as are the language experiments. However, Washoe is still a chasm away from the human infant so far as language potential is concerned.

Another important point is that children utilize abstract concept in a way that is quite beyond the capacities of Washoe. Dr. Mark P. Cosgrove puts it very well:

Animal signs such as those hand signs of Washoe are not true concepts, since a physical representative is always involved. A true concept, such as the meaning of the words "political science" could not be grasped by Washoe through any physical representation. One must not mistake animal knowledge in terms of visual experiences and concrete imagery with true concepts.[10]

This author (Wilson) supervised an extensive investigation into the acquisition and development of concepts by young children.[11] By five years of age the majority of the children could answer simple questions dealing with space and time, the measurement of heat, differences in size and weight, comparatives and superlatives, regular and irregular verb forms, pluralization, active and passive, past tense, and very much more.[12]

They had no special training, as Washoe had. They were, simply, normal children, with normal exposure to language. They were human: that made all the difference.

Footnotes: Chapter 3

1. Gardner, Drs. R. A. & B. T., *op. cit.,* Table 1, pp. 30-32.
2. *The Acquisition of Language,* p. 54.
3. Gardner, *op. cit.,* p. 28.
4. *Ibid.,* p. 28.
5. *Ibid.,* pp. 28, 29.
6. *Ibid.,* p. 33.
7. This terminology was introduced by Martin Braine in "The Ontogeny Of English Phrase Structure: The First Phase" in *Language,* Vol. XXXIX, 1963,

pp. 1-13. Other writers such as Brown and Fraser in "The Acquisition of Syntax" in Bellugi and Brown; and Muir and Ervin, "The Development of Grammar In Child Language," also in Bellugi and Brown, 1964, have continued the hypothesis. David McNeill has a good criticism of the concept in his article "The Creation of Language By Children" in *Child Language* pp. 356 ff.

8. In *Behavior of Non-Human Primates*, Vol. 3.
9. *Animal Behavior*, John Paul Scott, p. 225.
10. *The Essence of Human Nature*, p. 14.
11. This is well developed in unpublished thesis by Andrew Paton, "The Study of Relationship Between Semantic Development And Syntactical Development With Two Groups Of Children Aged 6 and 11 Years Respectively In Victorian State Schools," Australia.
12. An outstanding publication on concept formation in children was issued in 1944, the authors being Reichard S., Schneider M., and Rapaport D. The Title was "The Development Of Concept Formation In Children" and the article was published in the *American Journal of Orthopsychiatry*, Vol. 14, pp. 156-161. A number of later writers have drawn from this authoritative article. A selection of valuable writings in the area of concept formation would include:

Al-Issa, I. "The Development of Word Definition In Children" in *The Journal of Genetic Psychology*, Vol. 14, pp. 25, 28, 1969.

Ebanechko, P. O. and McGuire, T. O., "The Dimension of Children's Meaning Space," *American Educational Research Journal*, Vol. 9, pp. 507-523, 1972.

Klausmeier, H. J., Ghatala, B. S., and Frayer, D. A., *Conceptual Learning And Development*, New York Academic Press Inc. 1974. This volume has a thorough elaboration of the development of concepts in human beings.

Chapter 4

SARAH AND HER PLASTIC SIGNS FOR "DUMB" HUMANS

Another relatively successful experiment in teaching "language" to a chimpanzee was by Ann James Premack and David Premack, with their chimpanzee Sarah. We shall outline Sarah's achievements before we come to a summary of conclusions.

USING COLORED PLASTIC SIGNS

This experiment also started in 1966, in a laboratory of the University of California at Santa Barbara. In 1969, the time of their article, "Teaching Language to an Ape,"[1] Sarah had a vocabulary of about 130 terms that she used, with a reliability of about 75 to 80 percent.[2] As their medium of instruction the Premacks used colored pieces of plastic cut into various shapes to represent words. "Apple" was in the form of an equilateral triangle, "banana" was a square, while "apricot" was a rectangle. "Sarah" was in the form of a seated monkey as it would appear from the back, while "Mary" was a fancy large letter M. In some experiments actual objects were used, as when two apples and a banana were brought together and Sarah had to put the sign for "same" between the two apples and another sign for "different" between the apple and the banana.

SEARCHING FOR "UNDERLYING BRAIN MECHANISMS"

The Premacks make the point that they were using a system of naming that especially suited the chimpanzee, and in this way they hoped to find out more about its conceptual world. They state that "ultimately the benefit of language experiments with animals will be realized in an understanding of intelligence in terms, not of scores on tests, but of the underlying brain mechanisms."[3]

They accepted the premise that the first step in teaching "language" was to exploit whatever was already present, so they mapped out a program along the lines of what Sarah could already do, such as giving. They taught her the relationship between two individuals and one object — the donor and the recipient, and the object that was being transferred. As with Washoe, the system involved rewards, starting with a banana placed between the chimpanzee and her trainer, the banana being available for the taking. After this transaction had become routine a pink plastic square was placed in such a position that Sarah had to put it on a language board before she could get the banana.

This routine was then changed, with an apple replacing the banana, and eventually Sarah understood the meaning of the plastic signal for "give." If she mistook signs such as "give" she did not get what she wanted: as when she wrongly used the sign for "wash" — her apple was promptly placed in a bowl of water and washed. In this way she learned the specific actions associated with particular pieces of plastic board. Eventually all this was transferred to the actual manipulation of the signs themselves for such expressions as "give apple," and incorrect sequences were not accepted: she would gain the apple if she put out the sign for "give apple," but not if she put out "apple give."

At first sight this gives the impression that Sarah was using syntax, but in fact she was simply showing that she could put two signs together in a sequence that would bring a reward. This is nothing to do with syntax as such, but is simply a form of problem-solving. As Sarah (and later chimpanzees) developed these "sentences," this same pattern was followed. They had no idea of syntax as such, but simply utilized an increasingly sophisticated system in order to gain the rewards available for so doing.

LANGUAGE ACCORDING TO RULES

Here we have another difference from the acquisition of language by young children. As we have already stated, the young child simply does not start with verbs as action words, and also when the child reaches the stage of using two words for a sentence, he will not normally use the incorrect order. He will not say "apple give," but "give apple." He will not need to be deprived of his reward until he uses the words in proper order, for he does it automatically. It seems that he has an inherent capacity that causes him to utilize words according to clearly defined rules of language.

This is not to deny that there are remarkable achievements by Sarah the chimpanzee, as in her use of symbols to portray concepts such as "same" and "different." She was taught to use the symbol for "name of," and for "not name of," and she used them correctly to show the relationship between two other symbols for words. She could do this even for the objects themselves when the objects, not the symbols, were used. Thus she would put the symbol for "apple" and the symbol for "same as" next to an actual apple, or she would put the symbol for "banana", the symbol for "not," the symbol for "same as," and then an actual apple. Undoubtedly these were remarkable achievements which showed that this chimpanzee had mastered some forms of expression that had not been demonstrated by any other chimpanzee. She was even able to identify certain colors, sizes, and a crude form of "if . . . . then." This conditional relationship was taught to Sarah as a single word, with a plastic symbol that she had to place between two sentences.

The communication maze had become increasingly sophisticated, but it was still reward-reinforced behavior. As the Premacks state, "Sarah had to pay attention to the meaning of both sentences very closely in order to make the choice that would give her a reward."[4] The sentences were still learned in a relationship of one symbol to one concept, and the symbols themselves comprised a series of separate relationships. Sarah's achievement was at a high standard, but, as we have stated above, it was still in the category of complicated maze patterns or other reward patterns that have been utilized with various animals.

An example of one of the sentences referred to, demonstrating this concept of conditional relationships, is relevant: "Sarah take banana if/then Mary no give chocolate Sarah". Another was, "Sarah take apple if-/then Mary give chocolate Sarah." The fact is, Sarah was wanting to get the chocolate that she prized very dearly. So it was that laboriously, over a considerable period of time, she learned that if she took the banana she would not get the chocolate, but that if she took the apple she would get the chocolate. The reward would not be given unless the signs were arranged in their proper order. It was a complicated maze, but maze it still was, and a quite different process from the rule-oriented speech of even the very young child.

"A LONG CHAIN OF CONDITIONED REFLEXES"

George A. Miller quotes Pavlov as arguing that complex behaviors are built up automatically from simple reactions by the process of conditioning — "It is obvious that the different kinds of habits based on training, education, and discipline of any sort, are nothing but a long chain of conditioned reflexes."[5]

It is no new knowledge that monkeys are capable of a

high degree of problem-solving. As A. & D. Premack themselves state, "The important thing is to shape the language to fit the information-processing capacities of the chimpanzee. To a large extent teaching language to an animal is simply mapping out the conceptual structures the animal already possesses."[6] Sarah already had this capacity for problem-solving, for visual imitation, and for the imitative use of gestures. Those abilities were put to good use as she was taught to manipulate over 100 symbols.

It is interesting to notice in passing that Sarah was not taught to ask questions. The Premacks state, "Sarah was never put in a situation that might induce such interrogation, because for our purpose it was easier to teach Sarah to answer questions."[7] Possibly the asking of questions that demanded more than (e.g.) giving something a name, would involve reasoning and cognitive processes that are beyond the capacities of a chimpanzee.

As with Washoe, the Premacks make it clear that "at first Sarah made many errors, taking the wrong fruit and failing to get her beloved chocolate."[8] They tell us that after various strategies had failed, Sarah paid closer attention to the "sentences," and eventually she was able to show in proper sequence all the symbols that would ensure she got the chocolate.

All this resulted from reinforcement by rewards and is a quite different process from the virtually effortless acquisition of language by young children. Sarah could even learn such sentences as, "Green is on red if/then Sarah take banana," involving a change in the position of two colored cards (banana having replaced apple). It is doubtful if a young child would perform as well as Sarah did UNLESS the same strategies of withholding, patient repetition, and reward reinforcement were followed. What parent would apply such methods to their child? Nor would a parent need to do so. Once again, the human and animal processes are dramatically different.

THOSE SILLY HUMANS . . .!

Sarah performed well on this confusing task because of the forthcoming reward. Those silly humans would not give her what she wanted unless she followed a certain pattern and, though she failed at first, she paid greater attention, manipulated the plastic signs correctly, and the humans were true to their word and gave her the banana.

However, despite the rewards, she did not reach 100 per cent efficiency as we would expect with children who have learned something in language. We read such sentences as, "Again she performed at her usual level of accuracy."[9] The Premacks had already stated that this was between 75 and 80 per cent. On the same page we read, "Surprisingly often, however, she chose the wrong word."[10]

In this particular instance her trainers recognized eventually that the poor performance was probably due to the fact that Sarah was really expressing her preferences in fruit. This indicates that Sarah was apparently not reasoning in human language as such, with the words standing for the symbols that the trainers themselves had given. She had decided her own association for a particular symbol, and the trainers adapted their meaning to hers so that so long as she manipulated a series of signs correctly they would respond to her expressed preferences in fruit.

This might even indicate that Sarah was doing very well in teaching "language" to humans! Nevertheless her achievements, such as associating a symbol with the actual object it stood for, were quite remarkable. They show that chimpanzees apparently have a form of intelligence that has some similarity to the early sensorimotor intelligence of a young child. This is a term put forward by the Swiss psychologist Jean Piaget and refers to the early coordination by the young child of his sensory and motor abilities. Lessons learned from this teaching of "language" to a chimpanzee (with both

Washoe and Sarah for that matter) can be significant in some cases of deprivation of language in children, whether through brain damage, by accident, or by genetic upsets.

In discussing the two experiments undertaken with Washoe and Sarah, Professor Thorpe suggests that in some ways Sarah's achievement was superior to that of Washoe's, with a greater vocabulary, and "her use of it is in many respects more sophisticated and complex."[11] Professor Thorpe analyzes the 112 plastic "words" that Sarah had learned at that time and listed these as eight names of persons or chimpanzees, 21 verbs, 6 colors, 21 foods, 26 miscellaneous objects, and 30 "concepts such as adjectives and adverbs."[12]

He further points out that it is not possible to have an exact comparison with Washoe, because Sarah is not required to learn to *make* the words, but rather to use them. Sarah cannot make new words, though she can manipulate and maneuver the words that are available to her.

SEQUENCES DECIDED BY HUMANS

Once again, this is a dramatically different process from the acquisition of language by humans. Sarah could manipulate as many as seven symbols in a sequence decided by humans, in order to get her desired banana. Her potential combinations were very limited, and this is so with all the combinations she could make of her hundred plus symbols. The use of seven symbols is very good for a non-human subject, but by no means remarkable. It is well attested that some higher animals can utilize as many as eight "remembered" items at one time.

The human subject is not limited to 100 symbols or words that others create: he or she selects them from an almost infinite range by which he is surrounded. Even if he was limited to 100 words, the mathematical possibilities for rearrangement are staggering, involving a figure with a large number of zeros. There simply is no significant comparison as to the potentiality of the human subject when compared with a chimpanzee, even one trained with such method and diligence as Sarah.

Professor Thorpe quotes Dr. Premack as stating, "We feel justified in concluding that Sarah can understand some symmetrical and hierarchical sentence structures and is, therefore, competent to some degree in a sentence function of language."[13]

SEQUENCES THAT BRING REWARDS

With due recognition of the achievement in this case, we suggest that Sarah does not really grasp the "sentence function of language" except so far as the sentences consist of individual units. Those units are brought together one by one so that a reward that is placed at a progressively greater distance will eventually be gained. The rat learning its way through a maze discovers the right tracks by trial and error, and eventually finds the cheese at the end of the trail. That rat certainly did not know anything about the "sentence function of language," and, in the human sense of language usage, neither did Sarah. Her behavior was goal-oriented problem-solving of a sophisticated type.

Professor Thorpe suggests that another achievement of Sarah's is her ability to answer questions relating to her classification of particular objects, by the use of symbols for "same" or "different" and "not the same" or "not different." However, as Professor Thorpe himself points out, "There are indeed objections which can be made as to how far these are true questions."[14]

It is also interesting that Sarah appears to think of the sign as being the thing it represents and not simply as a piece of blue plastic. This is not surprising, for either way the same reward is forthcoming and the identification is complete.

This leads to another highly relevant point about Sarah's "language." As Dr. Mark P. Cosgrove puts it,

In spite of the fact that she mastered all the

operations required of her in the laboratory, Sarah's plastic "words" represent not true concepts, but only sensory abstractions. It is true that in sensation an organism will respond to a variety of relationships between objects. However, mere response to relationships does not indicate the presence of abstract thinking of them as relationships.[15]

In Chapter 3 we have already referred to the difference between human and infants at this point. As with Washoe, there is no evidence that Sarah has any capacity for abstract thought or true concept development. Such abilities are species-specific to humans.

Sarah's signs will continue to play their part, for the experiments have continued with other subjects. Hopefully, the chimpanzees will soon be able to get through to those dumb humans! There is little hope of ever getting humans to understand chimp chatter however, for one must be realistic in setting goals with investigations that involve such poor subjects!

Footnotes: Chapter 4

1. "Teaching Language To An Ape" Ann James Premack & David Premack, in *Scientific American*, Vol. 227 (4), Oct. 1972.
2. *Ibid.*, p. 92.
3. *Ibid.*, p. 95.
4. *Ibid.*, p. 96.
5. In *Psychology, The Science of Mental Life*, pp. 235-236.
6. Premack, *op. cit.*, p. 95.
7. *Ibid.*, p. 95.
8. *Ibid.*, p. 96.
9. *Ibid.*, p. 98.
10. *Ibid.*, p. 97.
11. "The Comparison of Vocal Communication in Animals and Man," W. H. Thorpe, p. 45 (In *Non-Verbal Communication*, Ed. Robt. A. Hinde).
12. *Ibid.*, pp. 45-46.
13. *Ibid.*, p. 46.
14. *Ibid.*, p. 46.
15. *The Essence of Human Nature*, p. 14.

Chapter 5
HOW'S YOUR YERKISH?

Viki's four words . . . . Washoe's use of American Sign Language . . . Sarah's sophisticated plastic talk . . . . what next?

A SENSATIONAL BREAKTHROUGH?

Experiments and investigations have continued, and the results in one center are, at first sight, startling. Then we look again and find that once again it is the same pattern of human patience, animals adopting problem-solving techniques, visual stimulation, tactile capacity, and reward reinforcement.

We shall briefly survey this very interesting investigation that at first seems to be a sensational breakthrough in the acquisition of language by chimpanzees.

A team of scientists have been working at the Yerkes Regional Primate Research Center in Atlanta, Georgia, with a chimpanzee called Lana and an orangutan named Biji. They set out with a four-year plan to teach these animals to communicate, and their methodology was to let them have whatever rewards they liked in the way of food, drink, music, films, and trips to the monkey playground, provided they correctly pressed lighted keys on a large keyboard naming appropriate symbols. In the first stages of the experiment, if the animal pressed one key correctly the desired reward would be available, but gradually each "trainee" had to combine a number of symbols before it could get what it was asking for.

"JUST AS A CHILD LEARNS" — ???

The Yerkes Newsletter[1] dated September, 1972, says they would be "expected to combine several different symbols, just as a child learns to form a sentence from words."

It should be said immediately, as stated in previous chapters, that this is not the way a child learns to form a sentence. Right from the very beginning of speech the child is giving a complete statement, and even when that child utilizes only one word it is an abbreviated "sentence," known technically as holophrastic speech. This demonstrates considerably more than the simple isolation of a particular symbol that is being utilized to draw attention to a specific object, with the intention of gaining a reward.

The project is headed by Dr. Duane Rumbaugh, and he acknowledges that the experiments conducted with the chimpanzee Viki, 40 years previously, showed that it was hopeless to attempt vocal communication with apes, for, as he says, "They just don't have the equipment."[2]

Rumbaugh has brought in other experts in linguistics and computer programming, and a computer system has been programmed so that the responses of the two apes can be carefully studied. The scientists hope for a breakthrough, with Lana and Biji demonstrating some sort of syntactic understanding in their requests. However, as we go on and study the reward reinforcement techniques utilized, it becomes clear that once again these apes have mastered increasingly complex puzzle-solving techniques. They utilize more and more words before they get the banana, the television entertainment, the opening of the window, the tickling, or whatever other desired reward for which they ask.

YERKISH LEARNED BY
REWARD REINFORCEMENT

The "language" that the apes are using is called "Yerkish," in honor of Robert N. Yerkes, after whom the center is named. The so-called language is made up of a series of "words" that are termed lexigrams, these being symbols on a keyboard which the ape can press. If the requests are made in a certain sequence, the reward follows. The apes soon learn that if a sequence is wrong, they do not get the reward.

Some researchers see this as indicating that these animals are utilizing a syntactic structure, but the fact is that the subjects learn by experience that they will get their reward only if they press the right levers for the particular word symbols — and in the correct order. If the order is wrong, they learn to start again. They also learn by experience that they cannot start again until the word "period" has been pressed, and so if they recognize they have made a mistake they press the symbol for "period," and they do start again. It is still a conditioned response with goal-oriented behavior.

All this is in the same category as the problem-solving techniques used with other animals and involving mazes and boxes. The desired rewards of bananas happen to be very suitable for chimpanzees, and they are prominent in the experiments. Nobody has ever doubted that monkeys are intelligent, and these types of experiments — that involve stimulus, response, and reward reinforcement — are not new. This is simply a new series of "tricks" that these apes are learning. In the process they are producing a form of communication that has some similarity to the simple aspects of human language.

The November, 1973, Yerkes Newsletter[3] shows that Lana then had mastery of 50 different symbols for lexigrams, and that from those symbols she could formulate a variety of requests as she asked for her food, liquid, etc. She could gain a great number of things simply by asking: "She is the only chimp in a serious research program who with malice aforethought is invited to spoil herself rotten in the interests of advancing science."[4]

LANA'S ERROR-CORRECTING TECHNIQUE AND OTHER REWARD-GAINING PROCESSES

After eight months Lana had made progress and had mastered several "sentences." She had also innovated the error-correcting technique of pressing the lever for "period" when she made a mistake. This was hailed as a significant step, but it is basically the same process that takes place when a rat returns to the beginning of the maze and starts again. Lana's reward was not forthcoming at the end of that particular "trail," and the way back to the "starting blocks" was via the "period" lever. It is not really surprising that she learned to press it when she made a mistake.

This however is not language in the true sense. Human language involves a large number of interrelated features. We shall see that there are many design features in language, as well as biological features, which make it impossible for any animal to participate in all the features that are requisites for pure language as such. Lana was communicating when she pressed that lever, but she was not merely saying to her trainers, "I have made a mistake." Rather she was responding to her conditioned cue to make sure she got that banana.

Despite the great effort, patience, and expense involved in this project, the researchers themselves acknowledge at the end of the summary, "Yet, at present there is no evidence which unequivocally supports the conclusion that they can acquire and use language."[5] "Language" in humans is very much more than reward-reinforced behavior.

The May, 1974, *Yerkes Newsletter*[6] gives a further progress report. At that time Lana was slightly more than three years of age, and she had completed a year in the computerized controlled language training situation. The claim is made that Lana had demonstrated ability "to communicate well beyond the demands of the chimpanzee's natural habitat in Africa."[7] However, Lana had learned to communicate only what she had been taught. If she had been sent back to her "natural habitat" in Africa, no doubt some of the things that other chimpanzees in the jungle knew would at first be foreign to her. In captivity Lana was simply showing that she could benefit from training.

Dr. Rumbaugh himself writes about the capacities of the chimpanzee: "It is known to be innovative in its tool use, hunting, and infant care. Even so, the natural haunts of the chimpanzee do not fully challenge its intellect."[8]

This is undoubtedly true, and with intensive training Lana was learning new methods of communication so that she could gain desired rewards. Just as the chimpanzee is innovative in tool use, so now Lana was innovative in learning how to achieve a wide series of pleasant ends.

LANA'S KEYBOARD AS A SOURCE OF REWARDS

By this time Lana's keyboard held some 75 keys, and the items to which the keys referred were animate, or physical, or things that could be eaten, or the keys could refer to various activities, or even to prepositions. The locations of the items were altered on the keys from time to time, and so it was clear to the researchers that Lana was recognizing the symbol itself, making use of visual stimulus. It was not merely memory learning as to where the particular keys were.

The availability of words that Lana could use was limited to the number of keys present and active on the console at one time. The number was usually about 50 words, these varying according to the particular aspect of the experiment being conducted. There is a very great contrast with humans: man's availability of words is limited only by his own memory bank and his extremely wide-ranging language environment, and the range of readily available words is vast. There is little comparison with the learning of words by a chimpanzee.

Dr. Rumbaugh also makes the point that Lana does not differentiate between the machine and the humans who attend her. She will tap out the symbols, "Please machine move into room" when there is no person present who can move into her room, or she might request, "Please machine tickle Lana:" in other words, the machine was a means of reward gaining. Dr. Rumbaugh himself said that Lana had equated the machine and the technicians "with regard to at least one dimension — they all fill requests."[9]

THAT WORD "PLEASE" — DID LANA LEARN POLITENESS?

At first sight it would even seem that Lana was taught politeness, for her requests started with the word "please." Actually there was a technical reason for this, for it was what the experimenters devised so they could signal to the computer that Lana was about to make a request. This was done by teaching her to depress the key for "please." In the same way she was taught that the signal for "period" was simply to tell the machine that the particular communication was finished.

The experimenters claimed they were introducing a form of holophrastic expression into Lana's vocabulary, and as an example they give the request, "Please machine give M and M period" — M and M being a candy. They liken this to the young child who says the one word, "Cookie," when she means "Mother, may I have a cookie?"

However, linguistically Lana's expression is very different from the child's. The child's one word "Cookie" is an abbreviated form of a complete sentence: it is itself a total unit, complete for that particular stage of the child's development. Lana's expression is entirely different, for there is no abbreviation at all. She has learned the one-by-one association of symbols, and if any one of them is missing she does not get the reward. The child uses one symbol as the expression of the whole, whereas the chimpanzee must use all symbols before the one result will be forthcoming. It is actually a contrast to the holophrastic expression of the child.

Undoubtedly Lana could press a lever for "Please," and she showed a great capacity to perceive and to follow, but the whole approach was quite different from that associated with children's language. At the appropriate maturational points the child's language is triggered by hearing language itself, in his or her environment. There is no reward reinforcement, as with the "spoiled" Lana, and the fact that the chimpanzee acquired increasing accuracy simply indicated that she became more and more confident with her ability to press the right keys in the right order so that the desired reward would be forthcoming.

Lana was also able to transfer the meaning from the signs she herself reproduced to those same signs as they appeared on the keyboard operated by the experimenters. This indicates that there was a visual understanding of the medium of communication.

She also "spontaneously" names such things as the food she eats,[10] reminding us somewhat of an incident with Washoe making the sign for "toothbrush" when she saw one in the Gardners' bathroom. Both animals were conditioned to respond in that way. Their "spontaneity" is not in the same category as children continually "labelling" new objects in their environment.

There are other experiments going on of a similar nature to those with Sarah, such as with Cheli, a female orangutan. At 19 months of age she was sent on five years' loan to Dr. David Premack of "Sarah" fame, at the University of California. Oglethorp is another chimpanzee, sent in October, 1973, to the communications project being conducted by the Gardners — the leaders of the team experimenting with Washoe. Thus the work with both Washoe and Sarah has been extended.

LANA AS A MUSICIAN?

The August, 1975, *Yerkes Newsletter* includes an article dealing with the possibility of Lana recording music, but again she is rewarded by candy when she presses a lever that will record a musical note.

Actually this tends to destroy the case for chimpanzees learning to "talk," for it highlights the nature of the whole process. Lana is undertaking reward-gaining performances of various types. If she really could speak and would express her mind, she would probably tell us that she did not mind humoring silly humans with their childish games, so long as they kept supplying those chocolate bars. Unfortunately they do not grow on trees known to Lana, and if the only way to get them is to learn Yerkish — well, there's plenty of time. Can you imagine Lana talking to a newcomer? "Let's get on with the job. Now see this symbol here on this funny machine — well, if you press this one, then that one, followed by . . . . you'll get a chocolate bar. Got it? Try it for yourself . . . . good girl!"

Footnotes: Chapter 5

1. *Yerkes Newsletter*, Vol. 9-2, p. 11.
2. *Ibid.*, p. 11.
3. *Ibid.*, Vol. 10-2, 1973.
4. *Ibid.*, p. 19.
5. *Ibid.*, p. 19.
6. *Ibid.*, Vol. 11-1, May, 1974.
7. *Ibid.*, p. 2.
8. *Ibid.*, p. 2.
9. *Ibid.*, p. 3.
10. *Ibid.*, p. 6.

PART II
THE BEES DANCE
TO TELL THEIR TALE

**A Survey Of Some Of the Major Communication
Patterns In The Animal World**

Chapter 6
WHAT IS COMMUNICATION?

No one doubts that animals "communicate," but there are differing opinions as to what "communication" is. One simple descriptive statement is that of Professor D. M. McKay of Keele University, England:

> A communicates with B only when A's action is goal-directed toward B . . . . and B perceives whatever he does about A.[1]

Because our purpose is nontechnical, we will not pursue the subject in technical terms. The fact that animals communicate is not in dispute, and in this book it is not communication that is being challenged: we are examining one major problem. We are asking ourselves if there is any possibility of nonhumans engaging in those forms of communication which have traditionally been recognized as unique to humans.

LEVELS OF COMMUNICATION

Various levels of communication have been suggested, and the following by W. N. Tavolga[2] has been well received:

1. Vegetable level, such as the possible effect of plants on each other, and also on those animals that feed on them.

2. Tonic level, relating to processes and functions such as food-finding and trail-following: excretion and cellular metabolism are typically involved.
3. Phasic level, referring to discontinuous stages that are more or less regular in the development of an organism. The emitting of the signal by one party would be discriminately received by the other: examples are sex signals in lower vertebrates and processes involved in the schooling of fish.
4. Signal level, involving specialized structures that produce a more specific, narrow band signal. This can involve biosocial signals, such as in the reproductive behavior of lower vertebrates and birds, and psychosocial signals, where the patterns are brought together in a relatively complex order, as with the song of a bird.
5. Symbolic level. This allows for considerable variation of the signal, and it is dependent on social interactions to be developed. The symbolic sounds and gestures are relatively primitive, and such activity is common with nonhuman anthropoids. Nonhuman primates make considerable use of gestures at the symbolic level.
6. Language level. This includes the "communication of abstract ideas, speech, meta language. [It is] restricted to man."[3]

Hebb, Lambert, & Tucker summarize a number of relevant propositions.[4] They point out that broadly speaking there are three levels of communication, these being

1. Reflexive, especially characteristic of infants but also found in higher animals and man.
2. Simple purposive gestures, which are common characteristics of dogs and apes, as well as humans.
3. Human language.

"A VERY LIMITED SET OF REFERENTS"

In his article "Messages of Vertebrate Communication," W. John Smith[5] makes the point that "in communicating, most birds and mammals appear to use a very limited set of referents."[6] He proposes a list of referents that could include most of the messages conveyed by nonhumans. The list is as follows:

1. Identification. This makes possible social interaction between classes, and "the most common classes are probably species, sex, maturity, and one individual."[7]
2. Probability. If one type of sound is used, certain other action is more probable than another form of action. If a greenbacked sparrow uses the "chuck notes" display, escape is more probable than attack, but "when the sparrow uses the 'medium hoarse notes' display, attack is as probable as escape."[8]
3. General set. These are associated with differing activities that range from such things as foraging, preening, and resting (these being maintenance activities), and going on to social activities.
4. Locomotion. Some displays are used only during locomotion, or at the beginning of locomotion, or at its end. The locomotion may relate to foraging, to patrolling, to following a mate or a parent, or avoiding being captured, etc.
5. Attack. This involves e.g. hostile acts and attempts to escape, or ambivalent behavior when either possibility is open.
6. Escape. This is when a communicator is e.g. either escaping or gradually withdrawing.
7. Non-agonistic subset is tied in with anxiety messages, or escape messages, etc. It can also involve associational contacts which do not include the possibility of the communicator attacking.
8. Association, as when one individual will associate with another but will neither come near nor allow the other to actually make contact. At other times it can be the passive allowing of another individual to approach.
9. Bond-limited subset. These are acts between

mates, or between parents and offspring, or members of large organized groups. There is a bond between the communicating parties.

10. Play. This applies to various nonaggressive contacts between bonded individuals, it often being a parody of non-play fighting among primates.
11. Copulation. These are displays used before or during copulation, by a number of species.
12. Frustration. This relates to certain behavior that would take place if the opportunity were available, or when that opportunity has passed.

LANGUAGE AS "SEMIOTICS"

Language is sometimes described as a system of signs, and as such it comes within the scope of "semiotics." Many authorities suggest that semiotics is broken up into syntactics, semantics, and pragmatics. Although there are variations between writers, the following statement of A. G. Smith would be accepted by many: "Syntactics studies how signs are added to one another. Semiotics studies how these signs are related to things. And pragmatics studies how they are related to people."[9] As John Lyons points out, this has some value for the study of other systems of signaling, but "is hopelessly vague for the study of human language."

Lyons lists some of the basic semiotic functions of language as follows:

1. Deixis, which draws attention to a particular object in the situation of an utterance as in the expression, "Look here!"[10]
2. Vocative signals, which attract the attention of a particular person, as in the expression, "Look out, John!" The person addressed is invited to assume the role of the receiver.
3. Nomination, where a name is a sign, as in the utterance, "This is John Smith."
4. Desiderative signals. This relates to the desire of the organism towards some object; it wants that object.

5. Instrumental signals, which serve to get something done.

CRITERIA FOR HUMAN LANGUAGE

In his introduction to *The Subject of Communication*, Robert B. Zajono[11] summarizes the six criteria laid down in 1929 by Bierens de Haan, the Dutch zoologist who set out to compare human and animal "language." According to de Haan, humans communicate by means of sounds that (a) are vocal; (b) are articulate, being composed of syllables joined into words; (c) have some conventional meaning — a meaning that is learned and is not given at birth; (d) indicate something: they do not only express emotions and feelings but they also refer to situations and objects; (e) are all uttered with the intention of communicating something to somebody else; and (f) can be joined into phrases in many new combinations.

De Haan concluded that animal "languages" meet only the first criterion more or less fully, and the second to a very small extent, but that in all the other criteria animal languages are deficient.

Obviously communication also involves a number of factors such as those listed by T. A. Sebeok: [12] source, channel, destination, code, message, and context.

A comparison of human language with the communication processes of all lower species is one way of highlighting the chasms between animal communication and human language. The mental processes at the source are incomparably inferior, at least according to standards applied by humans; the channels are simply not adapted for all that is involved in human speech; the destination is incapable of the variety, differentiation, and so much more that is taken for granted when the recipient is human; the codes are dramatically restricted with animal participants; the messages are necessarily highly restrictive and limited in e.g. social content, and they are without the rule-oriented syntactical potentialities available to humans; and the contexts are basically limited to immediacy and to self-interest.

ANIMAL COMMUNICATION CONTRASTED WITH HUMAN SPEECH

Undoubtedly many animals have elementary communication signals, but they are greatly different from the human speech system. Some of the commonly recognized patterns of animal communication include the following:

1. Animals have elementary communication signals with regard to e.g. pain and food, but these are rudimentary signals of limited dimensions, and they are basically emotional.
2. The bee's dance to show where food is, or the chimpanzee's cry, are very restricted as to internal structure.
3. There is no clearly-defined linear string of "words" with animal "speech."
4. There is no complex grammatical structure.
5. There is no great difference between geographically or culturally separated members of the animal community. On the contrary, humans separated geographically may need to learn another language.
6. Even though human language involves inherent factors, learning tasks are considerable — there is no counterpart with animals, and problem-solving techniques relating to intelligence are not entirely relevant: they are demonstrably unsuccessful with regard to human speech.
7. Physiological differences are very significant.
8. Compare cries of pain, fear, ejaculation, and onomatopoeia: although these are used in human speech, they are not the major factors of communication.
9. Animals have not the past/present/future potentials of human language.
10. They have very limited social communication.

SUPRASEGMENTAL FEATURES AND KINESICS

Suprasegmental features such as pitch, stress, tone,

and juncture can all make a dramatic difference to the meaning of actual words. A simple statement, such as "You do not want to go," will have different meaning according to which (if any) word is emphasized, or if "go" is said in such a way as to make the statement a question.

Human communication and human speech are not synonymous in other ways also. "Kinesics" or "body language" is also important. Humans can communicate by the raising of an eyebrow, the pointing of a finger, the stamping of a foot, and in many other ways, without actual words.

The complexity of human communication puts it on an infinitely higher plane than that known in the animal world.

Footnotes: Chapter 6

1. "Formal Analysis of Communicative Processes," in *Non-Verbal Communication* (ed. Robert A. Hine) p. 25. This informative article contains relevant technical data on communication processes.
2. In *Development and Evolution of Behavior*, L. Aronson, et al., 1970, Freeman, New York and San Francisco. Elaborated by R. A. Hinde in *Non-Verbal Communication*, pp. 92-93.
3. *Ibid.*, p. 93.
4. In "A D.M.Z. In The Language War," D. O. Hebb, W. E. Lambert, and G. Richard Tucker (reprint from *Psychology Today*, April, 1973, pp. 55-62).
5. In *Readings In Animal Behavior*, Thos. E. McGill (Ed.) Holt, Rinehart, and Winston, N.Y., 1973.
6. *Ibid.*, p. 295. Smith wrote as Associate Professor of Biology at the University of Pennsylvania, Philadelphia.
7. *Ibid.*, p. 297.
8. *Ibid.*, p. 298.
9. Smith, A. G., *Communication and Culture: Readings in the Codes of Human Interaction*, pp. 70-75.
10. *Ibid.*, pp. 70-75.
11. In *Animal Social Psychology*, p. 177.
12. "Semiotics and Ethology," in Sebeok, T. A. & Ramsay A. (eds.), 1969, *Approaches to Animal Communication*.

Chapter 7
THE BEES DANCE AND THE BIRDS SING

Communication takes place in many other ways besides "speech" as such. Even worms engage in communication for reproductive purposes, and the signals can be visual, tactile, and chemical.[1] Frogs and toads have mating calls which are markedly "vocal," though they are limited to four sounds that can be separately distinguished. Alligators roar, and this is a territorial defense device: it is common in the animal world for animals to mark out their territory, often by urinating, and intrusion will meet with resistance.

Some fish even "use electric pulses for orientation," and researchers believe that probably electrical discharges from fish have a social function: social function is prominent with other creatures. When we come to honey bees, we have what is widely accepted as a high order of social communication.

THE DANCE OF THE BEES

Karl von Frisch is renowned for his patient studies of bees and the communication involved in the "waggle dance."[2] By this method one bee can communicate to others where food is, both as to its direction and the approximate distance. The direction involves the bee being so many degrees away from the sun, and the distance involves the number of waggles of the wings per second. If the food was 200 meters (218.72 yards) away, the bee would be likely to waggle about eight times in 15 seconds, while at 500 meters (546.8 yards) it would waggle only six times. Beyond 5,000 meters (3.107 miles) the curve of the waggle dance would be practically level, and this is about the greatest distance to which any bee would go looking for food.

Even if only a small patch of blue is visible the dance can still be successful, for the degree of polarization of light is related to the position of the sun in the sky. Bees have an even greater advantage than humans in this respect, for if the sky is completely cloudy, so long as enough ultraviolet light can penetrate the clouds, they still give their message: they can see the sun even if it is invisible to humans.

In later studies than those of von Frisch, Adrian M. Wenner has shown that the bees could also transmit information by utilizing their odors and sounds, as well as dances. It is just as though there is an inbuilt computer inside these bees, making all these calculations and methods effective. Clearly this is a form of communication, for one bee is telling another where something else is to be located, even though it is out of sight of the communicator and the communicant, for the bees can read these messages from their companions much more easily than humans can.

It should also be noted in passing that Karl von Frisch's conclusions about the waggle dance have been challenged by Adrian M. Wenner and Dennis L. Johnson (see reference [2] above), claiming that their experiments showed that in fact bees would go to any one of four different sites, and not just to one as had been supposed. Other scholars suggest that the controls in their experiment were not sufficient, and that von Frisch's conclusions still stand. Richard Dawkins suggests that possibly bees have alternative ways of finding foods, including the dance, the smell, and also the presence of other bees. This being so it is possible that bees will complete their mission and reach the signalled food only if there are no significant intrusions. The "sidetracking" to other hives is therefore quite understandable: if a bee finds other good food on the way, why persist in completing a long journey to achieve the same end?

Professor Thorpe commented on the challenge by Wenner and his associates and acknowledged that they had added to our knowledge of the honey bee's communicative behavior, but argued that von Frisch's work was "overwhelming in detail."[3] Thorpe made the interesting comment that even a human observer could watch the bee's dance and predict both the direction and the distance of the food source — as he himself had found on a visit to von Frisch.

Chemical communication is also relevant in such creatures as ants. They can transmit an odor at eight-second intervals as a trail marker, thus pinpointing a food source close to the nest. Other chemical substances can be released as warning signals, and yet others as a means of sex communication.

BIRDS HAVING "CRITICAL STAGES" OF ACQUISITION

Birds develop their communication by song in recognizable stages. Chaffinches are at their maximum for learning at the beginning of spring, when they are about eight months old. The learning peaks at about 14 months, and soon after it ceases entirely.[4]

This strongly reminds us of the human capacity to acquire language in the early years of childhood, "cutting out" (in the sense of acquisition, not of formal learning) at the end of teens. For the birds to develop their song potential, "hearing must remain intact."[5] So long as that hearing pattern is unaffected in the early months of the bird's life, the song pattern will be clear, even if the bird is later deafened. This also has parallels with human speech, for the attainment of speech is very much more difficult for the child who has never heard.

Both with birds and with humans, auditory stimulus

and auditory feedback are each relevant in vocal communication. It is also relevant that birds have a surprising degree of individuality in their song patterns, and
this is important in such matters as the recognition of its
parent by a young bird.

Another point of similarity is that with birds the control for "speech" can be shifted from what is usually
regarded as the dominant side, to the opposite side.
However, this must be undertaken before song-learning
actually occurs.[6] After that it will not be successful,
reminding us somewhat of Professor Eric H. Lenneberg's
conclusions as to the restrictions concerning laterality
beyond teens with a human subject.[7]

The acquisition and learning pattern for the child is
consistent around the world, but it varies with different
species of birds. The pattern with the chaffinch is not
the same for all other birds, for with some "a period of
ability to learn new song elements may recur annually."[8]

Humans can continue to learn, not only annually, but
constantly. The difference is in the area of language
acquisition as distinguished from *learning*. The acquisition of language is a childhood phenomenon, with
restrictions on that acquisition as more formal stages of
learning are approached. The "acquisition" will normally be according to the social patterns and the language environment by which the child is surrounded.

BIRDS ACQUIRE A "DIALECT"

Not only humans have limitations on their acquisition of the forms of expression and communication
typical of their social group. That has been illustrated
by an investigation into the ways in which vocal behavior is transmitted in sparrows.

Two researchers of the Department of Zoology at the
University of California at Berkeley (Peter Marler and
Mirvako Tamura) reported such an investigation in
Readings in Animal Behavior. They showed that male
white-crowned sparrows acquire their particular song

"dialect" in about the first hundred days of their lives. This learning comes from hearing that "dialect" from older male birds. Once the hundred days have past, further exposure to other patterns is not effective. An "alien" dialect can be taught up to that time, but not thereafter. This is the same in principle as Professor Lenneberg's indications that human language should be acquired by about midteens.

However, there is a difference. The sparrow does not learn a second "dialect" after 100 days of life, but after midteens the human can learn another language. As we have said above, it will not be by simple exposure, as with the young child, but by applied learning. The difference from the bird is that the human *can* learn so long as he has acquired that first language.

The investigators also referred to a number of other patterns, including the need for the young bird to hear its own sounds if the "dialect" was to be acquired. They suggest that there are several aspects involved in the development of the song patterns of these birds, "including acoustical influences from the external environment, acoustical feedback from the bird's own vocalizations, and perhaps nonauditory feedback as well."[9]

"OTHER CRITICAL PERIODS" WITH ANIMALS

It is not really surprising to find that language acquisition takes place over a limited period of time, for with various animals there are crucial periods for particular skills to be acquired. Domestic chicks will follow a moving human being who happens to be present, as soon as they can walk. If this "imprinting" (as it is called) is allowed to develop, the chick will not even return to its own mother.[10] Although the imprinting could often be induced in some birds for several days, it was likely to be the case that it "gradually declined after the first few hours after hatching."[11]

There is a critical period for imprinting. With intensive application this can be extended, but not indefinitely. The onset and the conclusion of the critical periods are fixed by biological processes relating to growth and maturation. They can be extended to only a minor degree by such variables as "social facilitation," where there is a strong motivation to act according to the dictates of the social group.[12] Drugs such as meprobamate can also extend the period sometimes, but again for only the period that the metabolism has been slowed down, with a consequent stretching out of the critical period.

This process of "critical periods" is well-known in the animal kingdom. P. H. & M. S. Klopper[13] report a series of experiments relating to the maternal imprinting in goats. For a brief period after labor is completed, the mother goat is prepared to bestow her maternal affection even towards relatively old kids. Only a 5-minute period is required to establish an alien kid as being acceptable for adoption.[14] If her kid was stillborn, even her handler might find himself "overwhelmed by lavishly bestowed licks from a suddenly loving nanny."

It seems that even the instinct of fear in monkeys is innate and is induced by exposure to threatening stimuli. A threatening picture can apparently be sufficient to activate an "innate 'releasing stimulus' for fearful behavior. This innate mechanism appears to be maturational in nature,"[15] occurring between 60 to 80 days after birth.

In view of the many evidences for "critical periods" relating to various aspects of animal behavior, it is not surprising that humans also have critical periods for the acquisition of language. After those stages have passed, language development and learning are still possible, but not by the spontaneous processes associated with the acquisition of language by young children. The onset of puberty signals the "beginning of the end" for language *acquisition* as distinct from learning.[16]

Critical periods ... radar systems in bats ... the imprinting associated with various forms of animal life . . . the dance of the bees to indicate just where food is located . . . the navigational instincts of migratory birds — they all point to inherent abilities and to a consistent pattern of "design in nature." Human language is in the same category.

Footnotes: Chapter 7

1. The following examples are from "The Lower Vertebrates And The Invertebrates," by W. H. Thorpe in *Non-Verbal Communication*, (Ed. Robt. A. Hinde, pp. 127 ff.).
2. See "Honey Bees: Do They Use Direction and Distance Information Provided By Their Dances," Karl Von Frisch (pp. 310-326), and "Reply To Karl Von Frisch," Adrian M. Wenner and Dennis L. Johnson (pp. 325-327), in *Readings In Animal Behavior*, Holt, Rinehart, and Winston, N.Y., 1973.
3. Thorpe, *op. cit.*, p. 135.
4. *Ibid.*, p. 159.
5. *Ibid.*, p. 159.
6. *Ibid.*, p. 176.
7. Eric H. Lenneberg, *Biological Foundations of Language*, pp. 66-67, 181, etc.
8. Hinde, *op. cit.*, p. 159.
9. "Culturally Transmitted Patterns of Vocal Behavior in Sparrows" in *Readings In Animal Behavior*, p. 234.
10. Elaborated in "Crucial Periods in the Development of the Following Response in Young Nidifugous Birds" by Eric Fabricus, at pp. 237-247 of *Readings In Animal Behavior*.
11. *Ibid.*, p. 239.
12. *Ibid.*, p. 241.
13. In "Material 'Imprinting' in Goats: Fostering of Alien Young," P. H. & M. S. Klopfer, *Readings In Animal Behavior*, pp. 248-252.
14. *Ibid.*, p. 251.
15. In "Monkeys Reared in Isolation with Pictures as Visual Input: Evidence for an Innate Releasing Mechanism," Gene P. Sackett, in *Readings In Animal Behavior*, pp. 263-269.
16. Lenneberg, *op. cit.*, pp. 155, 156, etc.

Chapter 8
GESTURES OF GORILLAS AND SOUNDS OF DOLPHINS

Even in their natural state gorillas have well-developed capacities for communication, and remarkable tales have been told at times. Not all are in the realm of folklore.

LIVING WITH GORILLAS IN AFRICA

The zoologist George Schaller lived close to gorillas in African Congo (as it was then known) and so far as is reported was in more direct contact with these animals in their wild state than anyone before him. He found they lived in social groups of between five and 30, and that they had few natural enemies. Apparently they stayed together for companionship as much as anything. There seemed to be a head ape who led the other animals around, and they followed him more out of respect than fear.

In Schaller's book, *The Mountain Gorilla,*[1] he has drawn up what amounts to a dictionary of the communication sounds used by gorillas. They stretched from a purr of contentment to a grunt by which one animal would tell others in the tribe where he was. They had a warning "waw waw," a soft whine of distress, a playful chuckle, and a harsh "yoo-yoo-yoo" grunt whereby the males told the females to stop their quarrelling. A high-pitched scream warned the tribe of great danger.

In addition they had expressive facial features and could convey their feelings, telling when they meant to be playful, angry, curious, peaceful, afraid, or impatient. When individual gorillas were startled by Schaller coming upon them they would stand still and shake their heads, at times letting their mouths fall open at the same time. Schaller decided that this meant something like, "Peace, I do not mean you any harm." He himself used this gesture and was apparently accepted by the gorillas as a result. A stiff-legged walk was sometimes adopted by one of the leading male gorillas, and it seemed to be an order to the rest of the tribe to follow him.

"HELP ME CARRY THIS BOX!"

Chimpanzees have an even greater capacity to communicate by expression, and they appear to read the expressions of humans with whom they are in contact. Robert M. Yerkes has shown that chimpanzees will cooperate, as in sharing the carrying of a box that was too heavy for one of them. They would communicate by gestures, such as a begging motion. Sometimes they communicated their desires by pantomiming what it was they wanted their companion to do. They can communicate in other ways, as when they show a trainer where a sore spot is: dogs have been known to do similar things, even pushing their master's hands towards that sore spot.

Through the centuries similarities between the physical abilities of apes and men have been noticed. Both can manipulate objects with their hands and can walk upright, although humans do so much more easily than apes. Apes and men even endure many of the same illnesses and have similar diets.

B. Gilbert points to differences (as well as similarities), and he writes:

> However, one important difference should be mentioned since it apparently has great bearing on the question of communication.

Man and apes seem to have about equal
faculties for hearing and seeing. However, the
apes appear to give far more weight and pay
more attention to what they *see*, in comparison with what they *hear*, than men do.
Some observers believe that this difference can
be traced back to communication. Because of
speech, men have developed through practice
their power of discriminating between various
sounds to a greater degree than apes have.
Apes, on the other hand, are perhaps a bit
more sensitive to gestures, visual expression,
and signs than we are."[2]

Gilbert further states, "Other tests and observations
have shown that chimpanzees, like children, learn both
by imitation and by trial and error." He gives a number
of examples whereby one chimpanzee learned a particular trick such as opening a door catch, possibly almost by accident, but then he himself showed how it was
done, and the other apes also learned the trick immediately. They undoubtedly learned by experience. There
are well-known cases of chimpanzees using their intelligence to solve problems, as with the young chimpanzee Mammo who could not reach a banana in a tunnel,
so he made use of a nearby stick to push the banana out
through the other end.[3]

WHAT ABOUT THE DOLPHIN?

The differences between humans in their potential to
communicate with each other are relatively small when
compared with the differences between man and other
beings. Nevertheless, there are degrees of "oneness" between men and those beings. Men feel they have more in
common with a dog than with a fish, and yet there are
interesting arguments that possibly we have as much in
common with some water creatures as with dogs.
Whales, dolphins, and porpoises are not fish, but are air-
breathing water creatures — they are "cetacea."
These creatures are highly intelligent and are

emotionally sensitive. We human beings do not know a great deal about them, mainly because they live in a quite different environment from ours. Nevertheless through the centuries there have been strange stories about their high intelligence, and though some of these are in the nature of fable and old wives' tales, it is still true that there is remarkable "intelligent behavior" among these sea creatures. For instance, whales hunt in schools, driving fish before them, even as wolves cooperate to drive deer before them on the dry land. Whales show sympathy for each other, as when they return to support one of their number that has been harpooned. There are authenticated cases of such behavior.

Trained dolphins are well-known in various parts of the world, and as a result the so-called mythical stories are no longer treated as so much nonsense. There are records of how these fish can cooperate with fishermen in bringing in their catch, being rewarded by being given some of the fish and other victuals. Certainly they demonstrate intelligence as they come to understand man's desires and are prepared to cooperate, even though there is a reward motive in this cooperation.

There are even stories of fishermen whistling to the dolphins, which would then come and cooperate.[4] The relationship is not unlike that which some men have with dogs, and it is not fantastic to read that they help to harvest the sea's products. In coming days dolphins might well be used in this activity to a greater extent, as dogs have been used in land-based activities.

THEY ARE NOT ALL OLD WIVES' TALES!

As Gilbert says, "In general the old stories about the cleverness, cooperativeness, and amiability of the whales, particularly the dolphins, are proving, according to the observations of modern naturalists, to be anything but exaggerations. The sociability and intelligence of the dolphins seems to be even greater than was suggested by the so-called myths."[5] Sometimes there are several hundred in a school of dolphins, and it seems that they stay together to hunt, to defend themselves, and also to enjoy each other's companionship.

Although obviously dolphins cannot be tested for their "intelligence" in the same way that chimpanzees can because of the difference in their natural habitat, they have learned to push levers in order to secure food. They will respond to various instructions which they have learned to understand, even more quickly than chimpanzees. They have relatively good memories, and, like chimpanzees, they also learn through imitation, trial, and error.

Another good illustration of their intelligence is when a group was unable to get at an eel that was hidden under a rock. One of the dolphins went off and killed a scorpion fish having sharp poisonous spines, and then it brought the dead fish back to the eel's hiding place. He used it as a fob, forcing the eel to leave the rock under which it was hiding.[6]

Interestingly enough, it is clearly not only the size of brain that is relevant to intelligence. The sperm whale has the largest brain of any creature, weighing more than 20 pounds, as against 15 pounds for the brain of an elephant and less than a pound for a chimpanzee.

The bottle-nosed dolphin and man both have brains weighing between three and four pounds. In comparison with the size of the body, man is the "brainiest" of all creatures, and the dolphins are a close second. These are highly-intelligent creatures, and they communicate mainly through sound. They even rival noisy birds in their constant vocalization, especially in their various forms of whistle. These vary according to whether they are made above or under the water, with some of the sounds utilizing extremely high frequencies that are considerably beyond the range of the human ear.

DOLPHINS COMMUNICATE
BY SOUND SIGNALS

It is well-known that dolphins in captivity can readily respond to the voices of their keepers: we have said above that they themselves communicate by means of sound signals. This is somewhat different from chimpanzees, where the reception is more by the visual channel, and expression is more by gestures and the tactile modality than by ear and vocal communication. However, despite the high intelligence of these creatures, there is no evidence to suggest that their communication is other than by discrete symbols — not signals brought together in a continuous stream, as with a "human" sentence. They learn quickly, displaying an intelligence along the lines of what the Swiss psychologist, Jean Piaget, termed the sensorimotor level of young children. There is no evidence that dolphins have the complexity of language by which man is distinguished.

Though we might have reservations about some claims, there is even some reason to believe that dolphins can imitate "humanoid" sounds. In *Man and Dolphin* John C. Lilly claims that a dolphin quacked the letters T, R, R, after he himself had been using these in dictation. The dolphin could hear these sounds because of an underwater amplifying system Lilly had fitted up in his aquatic laboratory. He claims also that the same animal repeated the words "three hundred and twenty."[7]

As Gilbert points out, Lilly's tests "demonstrated that dolphins are among the great vocal mimics of the animal world."[8] They are capable of giving a large number of replies with sustained accuracy, but it is as mimics. The sounds they imitate are apparently meaningless so far as the dolphins are concerned.

John C. Lilly and Ellis Miller report on a series of vocal exchanges between dolphins under various conditions of control.[9] They refer to whistles, clicks, creakings, quacks, squawks, and blats. They suggest that these are related to different meanings, including distress, attention, and irritation. Noises are elicited by such stimuli as the spoor of other dolphins in the water, or by visual or acoustic stimuli associated with another dolphin, by stimuli of various sorts from humans, and by the presentation of toys and bubble-type objects.[10]

Where the dolphins are in close contact with each other their high-pitched sounds act as a form of underwater radar. Also, when they are able to have body contact, swimming together and mating, the male and the female are likely to emit various sounds for periods up to 20 minutes at a time as they play, court, and mate: "In a typical 24 hour day there is a total of at least four hours of vocalization, and on many days there is more than this."[11]

In various ways dolphins demonstrate a remarkable animal intelligence. The tricks they perform at various centers around the world, as well as their communication patterns, are further evidences that some nonhumans have capacities much greater than was thought by our grandfathers.

Investigations involving problem-solving by animals have become commonplace. Because such behavior is highly relevant to some of the fundamental issues of this book, we deal with it briefly in our next chapter.

Footnotes: Chapter 8

1. Schaller, G. B., *The Mountain Gorilla, Ecology, and Behavior,* University of Chicago Press, 1963.
2. *How Animals Communicate,* B. Gilbert, p. 123.
3. *Ibid.,* p. 126.
4. *Ibid.,* pp. 141, 142.
5. *Ibid.,* p. 142.
6. *Ibid.,* p. 146.
7. *Ibid.,* p. 154, (quoting from *Man & Dolphin,* John C. Lilly).
8. *Ibid.,* p. 156.
9. "Vocal Exchanges Between Dolphins," in *Animal Social Psychology.* (Robt. B. Zajonc, Ed.) N. Y., John Wiley & Sons, 1967.
10. *Ibid.,* p. 202.
11. *Ibid.,* p. 203.

Chapter 9
ANIMALS AND PROBLEM-SOLVING

Many experiments have indicated that animals have remarkable capacities for problem-solving. Pavlov's famous dogs helped psychologists to understand a great deal about conditioned reflexes, as when he showed that dogs would salivate at the ringing of a bell alone, after that bell had become associated with their food; dogs opening gates, rats finding their way through elaborate mazes, and ever so many other controlled experiments have demonstrated that problem-solving is certainly not limited to humans. Elaborate Pavlov-type experiments have continued with other investigators.

Thus W. J. Brogden conditioned a dog to associate a sound followed by a light.[1] Then he conditioned the dog to regard the light as an "avoidable response," and so it tried to escape when the light came on. Having established this condition, the next step was to expose the dog to the sound on its own, and sure enough it tried to escape. This demonstrated latent learning which goes beyond simple stimulus-response-reinforcement forms of learning.

THE REMARKABLE "CLEVER HANS"

One of the most remarkable examples of intelligence is probably not as widely known as some others: we refer to "Clever Hans," and his achievements are well described by Professor George A. Miller.[2]

An eccentric German named Von Osten set out to

show that animals were as clever as men, and he spent two years educating a horse which became known as Clever Hans. It would shake its head appropriately for "yes" or "no" and would communicate in other ways by tapping its foreleg on the ground. Clever Hans could apparently undertake quite difficult calculations involving the four fundamental rules of arithmetic. He could change common fractions into decimals and could then change them back again. He could even tap out what day of the month it was, could tell the time, and would shake his head to show that a mistake had been made if a musical chord was wrongly played on the piano.

Von Osten did not make money out of his project, and he allowed others to put questions to the horse. Eventually a commission of eminent zoologists and psychologists studied the animal and reported that all possibilities of deception had been completely ruled out. Thus the claims of Clever Hans' intelligence were accepted as being substantiated.

However, not everybody was satisfied, and one investigator named Oskar Pfungst initiated another series of tests. The questions to be asked Clever Hans were written on cards, then selected from a pile so that nobody but Clever Hans and the questioner knew what the question was. When this was tried the horse could not answer anything at all. Von Osten himself was not informed what the question was. Clever Hans would begin to tap out the answer, but would keep on tapping, all the time looking intently at the man who had asked him the question on the card, as though waiting for some signal to stop.

Pfungst realized that the horse was in fact looking for some small movement of the questioner's head, and he came to the conclusion that Clever Hans was not really giving the answer himself but was simply looking for an almost involuntary movement of the questioner's head. "After the last of the expected taps the person would relax ever so slightly, thus inadvertently and unconsciously making a tiny movement. This was what Clever Hans was waiting for."[3]

Miller goes on to show that once Pfungst had discovered this secret he himself could then get any answer he liked from Clever Hans. Thus it was not "the higher psychical faculty of reasoning, but the lower psychical faculty of perception, that made Clever Hans so remarkable."[4]

It is relevant to state in simple terms that psychologists generally accept that sensation leads to perception, which in turn leads to the forming of a concept. For a concept to be recognized there is a more complex process involved, for one perception must be associated with other perceptions, then they are brought together and a concept is grasped.

Thus Clever Hans' ability was in the area of perception rather than of conceptualizing. He was not capable of the involved reasoning processes common to humans, but for a reward such as sugar he could concentrate on his questioner and thereby demonstrate remarkable powers of perception.

As we study the achievements of animals, we find that they involve self-interest and are basically reward-oriented. As early as Professor E. L. Thorndike's famous experiments with cats, these principles have been recognized

THORNDIKE'S CATS AND MAZES

Professor Thorndike would place a hungry cat in a strong box, with one side open and food placed outside the box. The cat could see and actually sniff the food outside, but could not reach it. Bars prevented its escape, and it had to learn to operate a lever, and later two levers, before it could get at the food. With repeated attempts the cat would become proficient, and eventually it would go straight to the levers and operate them successfully.

It was Thorndike who initiated the idea of various animals learning to penetrate a maze. By trial and error they would learn which paths led to a dead end, and

eventually the animal would have no trouble in select-
ing the correct paths. Their goal-orientation (getting out
of the box) was sufficiently strong to motivate them to
persist until they found the way out. The solution to the
problem was accomplished by what has become known
as "instrumental conditioning." Self-interest and
reward-orientation led to a successful conclusion.

THE INTELLIGENCE OF ANIMALS

This is highly relevant in the study of animal achievements. Animals are capable of achievements involving
intelligence, in some ways not unlike that of humans in
the early months of childhood. However, animal
achievements do not include the more abstract reasoning of later years, or even the complex mental processes
of the four-year-old child.

We fully recognize the possibility of certain higher
animals approaching the intelligence of the younger
child in some areas of achievement, but we immediately make the further point that intelligence and language are not the same.[5] The fact that monkeys can
intelligently work out problems, or that rats are remarkable problem-solvers in their mazes, does not mean that
either of these species can, or ever would utilize human
speech and language as human infants do.

Many psychologists would not accept Thorndike's
conclusions relating to the transfer of his findings to
learning theory for humans. A cat would normally claw
its way out of trouble, and not go around pressing
strange levers. Apart from the given reward, it would not
understand the actual processes involved in lever pressing, and there would be no insightful learning involved.

Humans do not normally find the solution to a
problem from a random, chance activity such as the fortuitous pressing of a lever. A problem is first recognized
as such, and then it is resolved by reasoned steps. There
might be trial and error, but it will be part of a process
involving such things as observation, reflection, manipulation, and verbalization. The problem is analyzed

and then attacked, with varying success according to the complexity of the problem and the ability of the subject himself.

CHIMPANZEES JOINING STICKS

Wolfgang Kohler argued that true learning took place only when the responses to which the animal was conditioned came within the normal capabilities of that animal. He conducted a whole series of experiments with chimpanzees, and his conclusions are relevant to our present study of language learning by chimpanzees.

In one exercise the animal was chained to a tree, with a stick in reach. It played with the stick for a while, then discarded it. Then a banana was placed in view but out of reach. He tried to reach it with hands and feet, but eventually grabbed the stick and pulled the banana within reach. The animal had learned to use the available means to accomplish a desired end.

Kohler describes these experiments in *The Mentality of Apes*.[6] One situation required the animal to reach outside its cage and haul in a short stick, and then use the short stick to get a longer stick at some distance from the other end of the cage. Having pulled in the longer stick, it could then reach out to get the bunch of bananas some distance away from the cage.

This experiment was repeated with a number of chimpanzees, and they all followed the same behavior pattern — frustration, repeated attempts to tear down the bars, followed by a quieter surveying of the scene, and then working out the solution to the problem. To use a human expression, they used their brains.

Kohler even had chimpanzees fitting two short sticks together. They did this to make one stick long enough to pull in the fruit that was beyond their reach with only the short stick. The most direct route was not sufficient, so learning had to take place, with a reorganization that involved more than usual application on the part of the chimpanzee.

STORING CHIPS FOR A RAINY DAY?

Other investigators have demonstrated that chimpanzees have surprising abilities, especially where they are able to imitate their human trainers.[7] Problem-solving has been successfully attempted in many chimpanzee experiments.

One chimpanzee learned to obtain its food by inserting colored chips into a slot machine.[8] Having achieved this ability, it was then taught to get the chips themselves from another machine that utilized a different mechanism. The basic drive for food led to the learning of a related skill, and after a time the chimpanzee would undertake the second task independently of the first. It even concentrated on the colored chips that had the highest food-procuring potential and stored them against a rainy day that no doubt his mother had warned him about!

The fact is, chimpanzees (and other animals) are able to learn the separate segments of a particular task and then bring the parts together into a meaningful whole. Undoubtedly many animals exhibit "intelligent" behavior in the grasping of problem-solving techniques.

For the resolution of a problem with animals there should be a sufficient motivation: if they are hungry enough, they are more likely to resolve a problem if they know they will be rewarded with food. Human learning also has clear motivational patterns, but they are not so blatantly related to reward behavior as with chimpanzees and other animals.

SKINNER'S PECKING PIGEONS

Many other examples of problem-solving in animals could be cited, one of the most relevant in the study of language-learning being Professor B. F. Skinner's pecking pigeons. To that investigaton we now turn, for the controversy it engendered is highly relevant in the study of nonhumans attempting to learn skills usually associated with humans.

Footnotes: Chapter 9

1. Elaborated in "A D.M.Z. In The Language War" (D. O. Hebb, W. E. Lambert, and G. Richard Tucker) in *Psychology Today*, April, 1973, p. 57.
2. *Psychology, The Science of Mental Life*, pp. 235-236.
3. *Ibid.*, p. 236.
4. *Ibid.*, p. 236.
5. London, Routledge & Kegan, Paul, 1925; Pelican 1957.
6. Elaborated in *The Psychology of Thinking*, Robert Thomson, p. 40.
7. *Ibid.*, p. 158.

Chapter 10
PECKING PIGEONS ARE NOT BABBLING BABIES

In 1957 Professor B. F. Skinner of Harvard issued his book *Verbal Behavior*, and this was reviewed by Professor Noam Chomsky of the Massachusetts Institute of Technology.[1]

Skinner hypothesized that the environment must be recognized as the great conditioner. The term environment applied to all the conditions and stimuli by which a person is surrounded. He argued that all learning is dependent on a relatively few basic principles, and that language development is simply another facet of that universal system of learning: any living thing that can learn must learn in the same way that every other thing learns, and whatever *is* learned will be acquired in that same way.

In a series of experiments with pigeons he showed that by such stimuli as flashing lights and rewards for appropriate action, they could be taught relatively complex activities. Those pecking pigeons could even involve themselves in a form of pigeon ping pong! In the same way, the argument went, human language could be taught by imitation, stimulus-response-reinforcement, and the utilization of surrounding stimuli.

THE IMPORTANCE OF STIMULUS-RESPONSE-REINFORCEMENT

As a thorough-going behaviorist, Skinner regarded all

learning as dependent on stimulus-response-reinforcement processes, and he believed that man's only innate ability was his capacity to deal with stimulus-response situations. Language learning was, to Skinner, no different in principle from any other learning activity.

Thus verbal behavior was reduced to a simple stimulus-response-reinforcement process, eliminating what are usually recognized by linguists as intervening facets of language, such as meanings and the rules of grammar. Skinner's approach blithely overlooked the huge varieties of stimuli and responses that would be required if every separate sentence in human language really did depend on a series of separate stimuli.

Such an hypothesis is seriously impractical. Theoretically, there is no limit to the potential number of sentences that could be generated by normal humans. Nor could a child ever learn all possible combinations of sentences so that they would be readily called on when the appropriate stimuli were presented.

Against such a simplistic hypothesis Professor Noam Chomsky insisted that in practice the child grasped rule-principles that could be applied across the whole structure of the language by which he was surrounded. This made possible an efficiency which was inconceivable with the limitations set by a stimulus-response theory.

CHOMSKY'S FRONTAL ATTACK

Chomsky's forthright opposition was somewhat unexpected. He had published his book, *Syntactic Structures*, in 1957, but it was a linguistic textbook as such. His review of Skinner's *Verbal Behavior* (See reference [1] above) was a frontal attack on behaviorist psychology, and it went beyond linguistics into the realm of psychology. It soon led to bitter debate and controversy between opposing schools within both disciplines.

Chomsky argued that Skinner's claim that all verbal behavior was acquired (and was maintained in strength through reinforcement) was quite empty. To him, Skinner's notion "has no clear content, functioning only as a cover term for any factor, detectable or not, related to acquisition or maintenance of verbal behavior."

That is the relevance of this debate to this section of this book, dealing as it does with "Communication." There is no higher form of communication than "verbal behavior" (to use Chomsky's words), and the great debate on the nature of verbal behavior cannot be ignored in this context.

One point of controversy related to the original impulse towards language. Skinner hypothesized that verbal behavior resulted from stimulus-response associations, which to some extent depended upon another external organism if reinforcement was to take place. Sometimes the reinforcement would come through a type of sub-vocal conversation whereby the language-user would reinforce his own verbal behavior. According to this theory, there are multiple causes and great diversity involved in the re-combinations of language units.

Skinner would not accept concepts such as innate predispositions for language, and here he and Chomsky are diametrically opposed.

LANGUAGE IS CREATIVE

Chomsky insisted that the creativity of language must be considered, and that this cannot be accounted for by theories involving only stimulus-response-reinforcement. By five or six years of age "normal" children are producing a very large number of utterances, many of which they have never learned before, or even heard.

Chomsky strongly argued that a behavioristic learning theory such as Skinner's, involving habits, associations, and conditioning, could never explain the extremely complex human behavior that is involved in language acquisition and development. Behavioristic terms such as "disposition to respond" were unsatisfactory and lacked empirical justification. He recognized that stimulus and response are part of language development, but that there is very much more of a genetic nature to consider, and all too often behaviorists do not come to grips with the basic problem.

"EXTERNAL FACTORS . . . ARE OF OVERWHELMING IMPORTANCE"

Chomsky left no doubt as to his rejection of Skinner's philosophy and explanation of language acquisition and learning:

Skinner's thesis is that external factors consisting of present stimulation and the history of reinforcement (in particular the frequency, arrangement, and withholding of reinforcing stimuli) are of overwhelming importance, and that the general principles revealed in laboratory studies of these phenomena provide the basis for understanding the complexities of verbal behavior.[2]

According to Chomsky, Skinner's hypothesis is that the speaker's own contribution is "quite trivial and elementary," and that it would be possible to predict human verbal behavior by knowing a few external factors that he had already isolated with lower organisms. Chomsky is emphatic that Skinner's "astonishing claims are far from justified."[3]

Chomsky goes on to acknowledge that Skinner has undertaken an enormous range of research, but in a detailed analysis of the results he questions the relevance of these experiments on animals for human learning of language. Drive reduction and other motivations are found in lower organisms and also in humans, but humans are not limited to the motivations of those lower forms. Nor do humans necessarily have all the complex functions of some of those lower forms. At times separate organisms must be recognized as distinct from each other in their functioning.

Drive reduction and such motivations that are present

in lower organisms have not been empirically established as the basis for human language acquisition. Reinforcement alone is not sufficient to explain such learning: it simply is not true that children learn language "only through meticulous care on the part of adults who shape their verbal repertoire through careful differential reinforcement."[4]

Each child who learns a language in a sense constructs an extremely complex and abstract grammar for himself, in an astonishingly short time. Skinner's reinforcement theories are not sufficient to explain this phenomenon, and it would seem that all human beings have some specific inherent "data-handling" or "hypothesis-formulating" capacity, of unknown character and complexity.

Professor Chomsky argues that humans are uniquely endowed with a species-specific "language acquisition device." This is directly opposed to the behaviorist philosophy that there is no special inherent capacity to acquire a language, and that such acquisition is a matter of learning and training, of stimulus and response, with culture and environment the most important concepts for learning. At the opposite extreme to the behaviorists, Chomsky's so-called rationalist view specifies that the acquisition of language initially depends on an innate capacity — that children learn to talk because they have an inborn capacity for language learning.

On this view, the main role of the teacher is to help shape the particular linguistic competence which is the cultural heritage of the child, but the actual learning process itself will be minimal. There is some genetic transmission, not yet fully understood, for the language acquisition.[5] Chomsky calls this an "abstract language acquisition device" (LAD for short).[6] He suggests that such a concept is as acceptable as the fact of a nest-building instinct in birds.[7] Around the world there are many other well-known examples that could be cited besides that bird-building instinct. The migration of birds is another.

COMPUTERIZED PROGRAMS
IN BIRDS

There are many well-known cases of birds flying amazing distances without significant landmarks. The golden plover flies from Alaska across the Pacific to Hawaii, and apparently it charts its direction from the sun and the stars.

Experiments conducted with penguins in the Antarctic regions also indicated clearly that birds can guide themselves by the sun. John Paul Scott writes, "that the birds were guiding themselves by means of a sun compass was shown by the fact that on cloudy days their paths were less consistent. If the sun were completely obscured, the birds might wander in any direction. While the sun was shining, the birds moved in very straight lines."[8]

Scott suggests that this means they were constantly correcting their own paths according to the information they received from the sun, "with the aid of some sort of internal biological clock."

Other birds migrate by reference to the stars rather than the sun, being night birds — such as the indigo buntings. Experiments indicate that their skill for using the stars to navigate needs to be developed in early life, but the experiments do not explain how it is that the basic impetus to move south in the Northern autumn takes place "without cues from other birds."[9] Once again, it seems they have some internal computerized programming.

Such systems are "triggered" by environmental contact at the appropriate maturational point. The onset of air breathing in mammals is another example.

Thus it is reasonable to suggest that an "L.A.D." might very well be a species-specific endowment of humans. Certainly the experiments with the chimpanzees Gua, Viki, Washoe, Sarah, and Lana have not indicated that any other species have a capacity for speech as it is utilized by humans.

In a sense, they are extensions of Skinner's "pecking pigeons" experiments, though not directly related in time or space. The chimpanzees also depended on stimulus-response-reinforcement and imitation and reward-giving. They have demonstrated that chimpanzees cannot "talk" as children talk. The fact is, pecking pigeons and chattering chimps will never be babbling babies. They are not endowed with the appropriate genetic structure.

Footnotes: Chapter 10

1. In the journal *Language*, Vol. 35, 1959, pp. 26-58.
2. *Ibid.*, pp. 27-28.
3. *Ibid.*, p. 28.
4. *Ibid.*, p. 42.
5. The debate between empiricists and rationalists is well summarized by Vera P. John and Sarah Moskovits in "Language Acquisition and Development in Early Childhood," pp. 169-214 of *Linguistics in School Programs*, being the 1970 Year-Book of NSSE. In the same volume, see also "Language and Thinking," Richard W. Dettering, pp. 280 ff.
6. See e.g. "Symposium on Innate Ideas," in *The Philosophy of Language*, ed. by J. R. Searle, The Oxford Readings in Philosophy (London: Oxford University Press, 1971), p. 122.
7. "Recent Contributions to the Theory of Innate Ideas" *Synthesis*, XVII (1967), pp. 2-3.
8. *Animal Behavior*, (ed. John Paul Scott) p. 239.
9. *Ibid.*, p. 239.

PART III
THE SILENT APE
AND THE BABBLING BABE
A 23-POINT ANALYSIS

A Comparison And A Contrast In Relation
To The Principles Involved
In Language Acquisition By Children
And The Learning Of Communication Patterns
By Chimpanzees

Chapter 11
HOW CHIMPS
AND BABES
COMMUNICATE

POINT NO. 1: FOR CHIMPANZEES, "SPEECH" IS PRIMARILY MANUAL WHEREAS FOR HUMANS MANUAL COMMUNICATION IS SUBSIDIARY

"Among themselves, the chimps seem to communicate both by vocalizations and by gesture. The vocalizations are calls related to alarm, aggression, and sources of excitation."[1]

Undoubtedly the infant chimpanzee has much greater manual dexterity than the human infant. If necessary, the infant chimp hangs from the hairs of its mother's stomach within a very short time of its birth and is soon swinging from branch to branch in a way that even human adults could not emulate. Clearly, chimpanzees are "manually-oriented."

It was because earlier experiments with chimpanzees had demonstrated that these animals could not speak in the human sense that the Gardners hit on the idea of trying to teach them to "speak" by utilizing gestures and various forms of manual dexterity. The Gardners themselves are not deaf, and they did not know sign language at the beginning of their experiment. Their results have been remarkable, and we have seen that there has been some measure of two-way communication between adults and chimpanzees by utilizing manual "speech."

HOW HUMANS USE "MANUAL COMMUNICATION"

Undoubtedly "manual communication" is important for humans also, but it is secondary rather than primary. The quadraplegic, having no movement whatever of arms, hands, legs, etc., can yet communicate perfectly satisfactorily by vocal speech. Nevertheless we are not for a moment suggesting that manual communication is not important in interrelationships between humans. We saw that facial expressions themselves are highly relevant and so are movements of the body, such as winks, shrugs, nodding and shaking of the head, smiles, frowns, putting the hand to the forehead, lifting the eyes to heaven, and so much more. In these ways we communicate our feelings, our likes and dislikes, our preferences, and we reinforce what we have said in spoken words. Children point, wave, and use their bodies in various other ways as helps to verbal communication. Even the position of the body is relevant as we humans communicate attitudinal or emotional messages.

Such messages are very important in a teaching situation. Gerald G. Duffy discusses research that dramatizes the need for teachers to take note of kinesics, with special reference to the facial expressions and movements of the body. He states:

> The results of some recent research dramatize the need for teachers to take such a broad view of the language code. These studies indicate that the verbal part of a spoken message (the sounds, words, and sentences) has considerably less impact than a speaker's tone of voice and facial expression. In fact, the research indicates that only 7% of the impact

of a message can be attributed to verbal stimuli, while 38% is attributed to tone of voice and 55% to facial expression. With suprasegmental phenomes and kinesics playing such an important role in the communication of language messages, we certainly cannot neglect to provide instruction in these areas in our elementary school language arts programs.[2]

The reference to suprasegmental features needs comment. These include pitch, stress, and juncture. They help us to separate questions from statements, sarcasm from more direct expressions, and to understand what is meant when sounds are run together, as in "a nice box" or "an ice box."

Suprasegmental features and kinesics (including body movements) are clearly important in communication between humans and are part of our language code. However, as we have stated above, they are secondary or subsidiary, with vocal speech primary for humans.

CHIMPS' NATURAL COMMUNICATION

Eugene Linden has written at length about the training of Washoe and other chimpanzees, and in *Apes, Men, And Language* he states: "Among themselves, the chimps seem to communicate both by vocalizations and by gesture."[3] He further reminds us that (chimpanzee researcher) "Jane Goodall documented gestures associated with begging, anxiety, and urging an infant to mount its mother's back prior to flight."[4]

In that same context Linden states that "all the experiments previous to Washoe had assumed that language was synonymous with speech," and suggested that this assumption "implicitly excluded from the temple of language a sizable number of people who do not use spoken language — the deaf."[5] He states that the Gardners wondered whether "some gestural language might sidestep the chimp's obvious difficulties in controlling its vocal productions."[6] Linden quotes anthropologist Gordon Hewes as suggesting that before the development of spoken language, man made use of gestures as his language.[7] The argument is that "both motor skill sequences and sentence constructions are adversely affected by the same lesion in many instances."[8]

It is relevant to point out that the term "motor skill" involves much more than gesturing, and in fact Linden himself further quotes Hewes that "the visual, kinesthetic, and cognitive pathways employed in tool-making and tool-using coincide with those which would have been required for a gestural language system. Speech, on the other hand, utilizing the vocal-auditory channel, implied the surmounting of a neurological barrier . . . (namely, associating visual stimuli with sounds)."[9]

This quotation includes an acknowledgment of the fact that speech utilizes a different channel and that a neurological barrier must be surmounted. No one denies that man is able to communicate by gestures. However, when we consider the communication patterns of chimpanzees, there is a dramatic difference between that sort of communication with all its limitations and the virtually infinite capacity open to man by utilizing his vocal apparatus. As he goes beyond gestures and utilizes the vocal/auditory modalities, he leaves all animals far behind. The chimpanzee's "natural" language involves the physical and manual, e.g. "Standing erect is an aggressive posture; grinning expresses fear."[10] Human communication by speech is much more specific and detailed: gestures are primary in chimpanzee communication, but they are secondary in human language.

Thus there is a basic difference in the first essential of speech. Chimpanzees communicate by gestures and by manual signs. Humans use spoken words. There are other basic differences, and one of these is in the realm of imitation.

POINT NO. 2: THE IMPORTANCE OF IMITATION

The Doctors Gardner write, "The imitation of apes is proverbial."[11] They are very ready to engage in visually guided imitation, and for chimpanzees imitation in the auditory realm is almost nonexistent.

There are two ways in which imitation with human infants is different from that with chimpanzees. As Courtney Cazden puts it:

The commonsense view of how children learn to speak is that they imitate the language they hear around them. In a general way, this must be true. A child in an English-speaking home grows up to speak English, not French or Hindi or some language of his own. But in the fine details of the language-learning process, imitation cannot be the whole answer. Examples . . . . of children's creative but rule-governed "errors" in inflections, questions, and invented spellings attest to an active, non-imitative construction process.[12]

This highlights two different aspects of imitation. Children imitate, and so grow up in the particular linguistic culture with the accent of the group by whom they are surrounded. On the other hand, when the child says "mouses" instead of "mice," obviously he is not imitating the models of language by which he has been surrounded from birth. Rather he has heard a word such as "house" and knows that its plural is "houses," and so he generalizes that rule to the word "mouses."

Thus we have so far seen two aspects of communication that are both primary for chimpanzees, but are secondary for human infants. Those two aspects are:

1. Manual/gestural communication
2. Imitation

Under Point No. 3 we consider an important area where the roles of "primary" and "secondary" are totally reversed.

In the second sense, outlined above, imitation is but

secondary in the role of language learning and acquisition. Again quoting Cazden:

> At most, imitation guarantees that the child's language system will converge, in superficial forms, on the language of his speech community. But it cannot account for the child's acquisition of the system of which those forms are the external expression. In the case of *a* and *the*, for example, the child can learn those two forms by imitation, but he must learn by some other process the abstract contrast in meaning that underlies their appropriate use.[13]

In recent times it has been increasingly recognized that imitation is secondary, not primary, in human language acquisition.

POINT NO. 3: VOCAL COMMUNICATION IS NOT POSSIBLE FOR CHIMPANZEES

No animal will ever communicate by human speech to the extent that humans do. This is true for a number of reasons, touching on interrelated disciplines such as physiology, biology, and psychology. The literature is extensive, and as it is studied it becomes clear that there are great chasms between the most highly developed forms of animal communication and normal human speech. In this chapter we are considering only one chasm as we recognize the uniqueness of the human vocal chords.

Philip Lieberman of the University of Connecticut wrote an article entitled "Primate Vocalizations and Human Linguistic Ability," and it has been of sufficient importance to be separately reprinted from the *Journal of The Acoustical Society of America*.[14] In it he refers to data indicating "that the nonhuman primates would not be capable of producing human speech even if they had the requisite mental ability." He makes a number of relevant points, including the fact that "unlike

man, the nonhuman primates do *not* appear to change
the shape of their supralaryngeal vocal tracts by moving their tongues during the production of a cry."[15] He
further states in that same context that "the basis for
the nonhuman primates' lack of tongue mobility appears to be anatomical." Much of his paper is of a technical nature and should be consulted for specific details, such as the fact that "the tongues of the non-
human primates are long and flat, and their supralaryngeal vocal tracts cannot assume the range of shape
changes characteristic of human speech."

NOT AN "OVERLAID MECHANISM"

Lieberman demonstrates that human speech is not
simply a fortuitous "accident," something overlaid on a
mechanism that was originally useful only for eating and
breathing.

This argument of an "overlaid mechanism" has often
been assumed as basic to the development of human
speech. As Lieberman shows, it simply is not a valid
argument. Man's capacity for speech is unique. Other
species can imitate some sounds, but only man has the
sustained control of sound units that is essential to
meaningful human speech.

Nor is it valid to argue that manual signs can replace
the use of the vocal chords as an acceptable alternative
for human communication by spoken words. There is
convincing evidence that the various forms of manual
expression utilized by the deaf have significant differences from human communication by spoken words.
This is perfectly understandable in view of "their
attachment to observed data."[16]

Normal human speech/language is necessarily restricted when auditory rehearsal is nonpresent or is
minimal. If speech were merely an "accident" of
development, then apes and monkeys would possibly
speak better than humans, for they have better breathing systems than humans do.[17] The simple fact is, they

are incapable of human speech with all its fineness of control and coordination.

Apes and monkeys do not involve their tongues in making noises, and for the first weeks of their lives human infants do not either. However, by the sixth week children have begun "to change the configurations of their supralaryngeal vocal tracts during a vocalization. The nonhuman primates never reach this point, though their general mental ability and physical dexterity are equivalent to, or better than, the human infants at this age."[18]

In this chapter we are mainly concerned with the vocal aspects of human communication. The "design features" outlined by the structural linguist C. F. Hockett are also relevant, but are outside our present scope except to outline them briefly. He elaborated 13 features in 1960, but in association with S. A. Altmann they were later expanded to 16.[19] The 16 "design features" deal with such aspects as the use of the vocal-auditory channels, rapid fading of speech signals, the displacement from time and space, and reflexiveness, by which is meant the ability to talk about the communication system. Separately, they are not all unique to man. Other species might have several of them, especially in a rudimentary form, but only man has the 16 design features to the highly developed form found in man.

Briefly, they are as follows:

1. Vocal-auditory channel.
2. Broadcast transmission and directional reception.
3. Rapid fading of the speech sounds.
4. Interchangeability — receiving and transmitting signals.
5. Complete feedback of own voice.
6. Specialization leading to effortless production of speech sounds.
7. Semanticity.
8. Arbitrariness — The tie to meaning is not linked to a physical or spatial relationship (e.g. "seven" has

more sounds than "nine" but its intrinsic meaning is less).
9. Discreteness, e.g. similar words are yet separate units.
10. Displacement in time or space.
11. Openness (or creative productivity).
12. Tradition (can be passed on).
13. Duality of patterning — Separate sounds become meaningful when brought together in a pattern.
14. Prevarication.
15. Reflectiveness (sometimes called reflexiveness — the ability to talk about the communication system itself).
16. Learnability.

All 16 are required for communication by human speech, and the total is to be found only in man. Speech/language is still uniquely the prerogative of man. The same conclusion is demonstrated by Otto Koehler's 19 features demonstrating man's biological uniqueness.

Animals communicate in various ways, but true speech is restricted to man. Undoubtedly chimpanzees would be among the closest to man in their capacity for communication, but sustained communication by spoken words poses an impassable barrier for any chimpanzee.

"THEY JUST DON'T HAVE THE EQUIPMENT"

We have seen that one of the projects involving chimpanzees was at the Yerkes Regional Primate Research Center in Atlanta, Georgia, with the chimpanzee Lana. That project is headed by Dr. Duane Rumbaugh, and he openly states that the earlier experiments with the chimpanzee Viki, 40 years previously, had shown it was out of the question to attempt vocal communication with apes, for "they just don't have the equipment."[20] Another relevant comment about experiments with chimpanzees is by Professor Roger Brown of Harvard:

> The chimpanzee articulatory apparatus is
> quite different from the human, and chim-
> panzees do not make many speech-like
> sounds, either spontaneously or imitatively.
> It is possible, therefore, that Viki and Gua
> failed not because of an incapacity that is es-
> sentially linguistic but because of a motoric in-
> eptitude that is only incidentally linguistic.[21]

One of the early attempts to teach apes to speak was
by William Furness in 1916, and he taught an ape to say
"papa" and "cup." Eugene Linden tells us, "Furness
noted that neither the orangutan nor the chimp seems to
use lips or tongue in making his natural calls; neither of
the words the orangutan learned required precise control over those physiological features."[22] Linden puts the
position succinctly when he states, "We would never expect the chimp to speak with the facility of a person,
who has been equipped for that task over several million years."[23] The reference to "several million years" is
totally conjectural, and there is no convincing evidence
to suggest that even if given "several million years" the
chimpanzees would speak as humans do. The total number of necessary "coincidences" would be overwhelming.

We have already discussed Viki, the chimpanzee
raised by Dr. & Mrs. Keith Hayes in Florida in the early
1950's. With a great deal of effort and by using her fingers on her lips in the way her trainers had done, Viki
eventually could produce sounds approximating "up,"
"cup," "mama," and "papa." We again quote Eugene
Linden, if only because he is not postulating the view-
point of this author (Wilson): we want to make it clear
that the problems are recognized by others besides those
who argue for impassable chasms between chimpanzees
and humans. Linden states, "Although Viki drove herself into paroxysms of frustration attempting to mimic
words, she was never able to reproduce more than a few

with any regularity."[24] Linden goes on to state that although she could form these words with her mouth, "For some reason she could not properly control her supralaryngeal tract." This again highlights the points referred to above by Philip Lieberman. Chimpanzees will never utilize human speech as a primary medium of communication.

MAN'S SPECIAL EQUIPMENT LEADS TO SUPERIOR PERFORMANCES IN SPEECH

For the human, speech is primarily vocal, utilizing the vocal chords, the tongue, the lips, the palate, the teeth, and so much more. Another whose contribution is highly relevant is Professor Eric H. Lenneberg in his classic text *Biological Foundations of Language.* The following are just a few relevant points:

The orangutan's . . . vocal organ is not capable of producing delicately modulated or controlled sounds. [25]

The chimpanzee also has a double set of vocal folds, but in contrast to both gibbons and man he can articulate each set independently.[26]

The chimpanzee has well-developed aryepiglottic folds which allow him to vocalize not only upon expiration but also upon inspiration, which for man is strenuous and unpleasant.[27]

It is interesting to note that in a certain sense man's vocal apparatus is in several respects simpler than that of the great apes. The geometry of the air spaces and fixed resonance chambers is "streamlined"; there is only one set of functional vocal chords; the vocal chords are mounted in the air tunnel in such a way that, when adducted, they can produce sound only (or primarily) on expiration, instead of allowing both inspiratory and expiratory voice; and the epiglottis has moved so far below the pharynx as to allow the air from the larynx to stream freely through both nasal and oral cavities.[28]

These facts again make it clear that though man communicates by speech, in some ways his associated equipment is inferior to that of the chimpanzees and other apes. If it be argued that deficient physiology is the reason why the chimpanzee does not acquire human speech in the full sense that the human infant does, further arguments from Lenneberg become relevant. He discusses the cases of children who for various reasons have severe articulatory deficiencies. He states that "these deficiencies need not impair capacity of the children for language acquisition."[29]

Lenneberg points out that this emphasizes that the brain itself is of primary importance during the time that language is in the process of development, rather than peripheral phenomena associated with articulatory deficiencies. Adults who have undergone oral pathology, such as amputation of the tip of the tongue, or those having a hairlip, "manage nevertheless to produce sounds that are fair approximations to those of normal speech, thus illustrating the same point."[30] He contrasts this with the ability of apes:

This ability contrasts markedly with that of apes with normal oral configurations and good homologies in structure. Their inability to approximate human sounds would, again, clearly indicate neurophysiological differences that are primary to the anatomical ones.[31]

SPECIES-SPECIFIC BEHAVIOR PATTERNS

Even the human who has various disadvantages acquires speech: "It is as if the machinery as a whole were put together so well that certain species-specific behavior patterns persist in an individual, even if certain supporting or coordinated mechanisms are destroyed or have failed to develop."[32] Lenneberg puts the situation as to man's uniqueness very clearly by his statement, "The alterations of the vocal tract that are specific to man can hardly be explained as adjustments to a different diet or as peculiar adaptations to any other vital function."[33] We emphasize that Professor Lenneberg is not accepting the philosophy of this present writer, and in fact he leaves unanswered questions as to the actual origin of language in man as distinct from chimpanzees and other animals.

Another point made by Lenneberg is also highly relevant in the comparison and contrast of the linguistic potential of chimpanzees and humans: we refer to developmental histories. Lenneberg states, "Maturation is a much more dramatic event in the human brain during childhood than in the chimpanzee's brain during the comparable period."[34] In the same context he further states that "man's brain maturational history is unique among primates . . . all lower forms approach the adult condition at a relatively quicker pace than man."

There are important differences in the developments between the brains of humans and chimpanzees:

> By extrapolation, we may assume that the maturational events of the chimpanzee brain during childhood differ from those in man in that at birth his brain is probably much more mature and all parameters are probably more stabilized than in man. This would indicate that the facilitation for language learning is not only tied to a state of flux but to a maturational history that is characteristic for man alone.[35]

(Maturation is touched on several times in this presentation — see e.g. Point No. 23 of this analysis.)

LANGUAGE LEARNING BY DEAF HUMANS

As we consider the implications of the above quotations, it is clear that, for whatever reason, chimpanzees cannot "talk" in the sense that human infants can. It is not sufficient to say that deaf people are also barred from human speech, for with intensive training they can in fact utilize speech in much the same way as other humans who are not so disadvantaged. It is this author's privilege to work professionally with a profoundly deaf colleague. He was so born, but despite that serious handicap he is highly qualified in more than one area, including a Ph.D. from one of the world's best known English Universities. His disadvantage was known at an early age, and he had special training from childhood. Today his lip reading is astounding, and his actual speech is perfectly intelligible. He is a living demonstration of the fact that speech in humans is far more than the mere ability to coordinate motor functions with various neurological patterns.

Humans are created with an innate capacity for communication by vocal speech. Chimpanzees are not endowed in this way. The chimpanzee Viki was terribly frustrated because it could not form words, many of which were easily formed by the child with whom it was reared. Experiments to teach chimpanzees actual speech have been dismal failures.

The fact is, human speech is not only "learned" but is acquired by the human child who is exposed to language and interchanges with language. Language acquisition is a specific ability in humans, related to, yet separate from, other learning abilities. The chimpanzee can "learn" certain skills, and can even learn to adapt certain sounds so that they will approximate relatively easy words such as "up" and "cup." However, the frustration that Viki showed is because she cannot cross the barriers: she cannot acquire language, for she is not so created. In comparison, the deaf child *can* acquire human language (including speech) when the particular problem is recognized and appropriate steps are taken for remediation.

SWIMMING IN AN OCEAN OF LANGUAGE

An interesting parallel to language acquisition comes from the realm of swimming. Many humans cannot swim and would die if they were thrown into deep water. Others can "dog-paddle" and would be able to save themselves provided they did not have to propel themselves too far to the shore. Others are highly proficient, even reaching Olympic standard. Yet the most proficient human swimmer would never suggest that he was at home in the water to the extent that a shark or even a porpoise was. No one would argue the point that they are not equipped for water living, as the inhabitants of the sea are equipped. For whatever reason, man does not have gills, fins, or the tail of a fish. He learns to be proficient in the water, but it is not his native element.

As we say, there is some similarity to the chimpanzee who breaks into the realm of human speech. He might learn to say "mama" and "papa" and might even become proficient in the use of two or three hundred plastic signs, but his achievement will be dramatically different from any normal child of five years of age. That child is surrounded by a great variety of language symbols and speech units — he is virtually swimming in an ocean of language. Yet in an amazingly short time he is striking out, surprising his delighted parents as he rides the speech waves so confidently. He is not drowned in an impossible situation, but keeps coming up for more. That is part of his human heritage.

If it be argued that it is only a matter of degree, then let it also be stated that the degree is rather like the difference between climbing Mount Waverley where this author resides (perhaps 800 feet above sea level), and climbing the mountains in the Himalayas, going as they do to some 30,000 feet. Men swim in water, but they are not fish; chimpanzees can make some limited use of human words, but they are still chimpanzees. They are certainly not human beings who happen to have tails. Chimpanzees have animal tails but will never tell human tales.

Footnotes: Chapter 11

1. Linden, Eugene, *Apes, Men, And Language*, p. 12.
2. Duffy, Gerald G., *Teaching Linguistics*, p. 33.
3. Linden, *op. cit.*, p. 12.

4. *Ibid.*, p. 16.
5. *Ibid.*, p. 16.
6. *Ibid.*, p. 16.
7. *Ibid.*, p. 62.
8. *Ibid.*, p. 63 (Quoted from Gordon Hughes, "Primate Communication And The Gestural Origin Of Language," in *Current Anthropology,* Vol. 14, No.1, (1973), University of Chicago Press.
9. *Ibid.*, p. 70.
10. *Ibid.*, p. 83.
11. In *Language In Thinking,* p. 19.
12. *Child Language And Acquisition,* p. 91.
13. *Ibid.*, p. 93.
14. *Journal of the Acoustical Society of America,* Vol. 44, No. 6, December, 1968, pp. 1574-1584.
15. *Ibid.*, p. 1580.
16. Olérsn, P., "Conceptual Thinking of the Deaf," in *Language in Thinking* (Ed. Parveen Adams), p. 48.
17. Lieberman, *op. cit.*, p. 1581.
18. *Ibid.*, p. 1584.
19. Linden, *op. cit.*, p. 83.
20. *Yerkes Newsletter*, Vol. 9-2, September, 1972, p. 11.
21. In *Psycholinguistics,* p. 210.
22. Linden, *op. cit.*, p. 14.
23. *Ibid.*, p. 122.
24. *Ibid.*, p. 244.
25. *Biological Foundations Of Language,* p. 47.
26. *Ibid.*, p. 47.
27. *Ibid.*, p. 47.
28. *Ibid.*, pp. 47, 48.
29. *Ibid.*, p. 51.
30. *Ibid.*, p. 51.
31. *Ibid.*, p. 51.
32. *Ibid.*, pp. 51, 52.
33. *Ibid.*, p. 71.
34. *Ibid.*, p. 170.
35. *Ibid.*, p. 174.

Chapter 12
WHAT ABOUT EYES AND EARS?

We have considered the role of gestures and other manual abilities. We have also recognized the importance of the vocal chords for the production of the speech sounds utilized in human language. In this chapter we analyze the importance of visual and auditory stimuli respectively, then we go on to other related points where chimp "talk" and human language are dramatically different.

POINT NO. 4: THE ROLE OF VISUAL STIMULUS

We have seen that Drs. R. A. & B. Gardner tell us that in their research they have found that chimpanzees are ready "to engage in visually guided imitation."[1] They quote Dr. Yerkes of the Yerkes Institute as writing, in relation to the two chimpanzees Chim and Panzee, that "their imitative tendency is as remarkable for its specialization and limitations as for its strength. It seems to be controlled chiefly by visual stimuli. Things which are seen tend to be imitated or reproduced."[2] The quote goes on to say that what is heard is not reproduced, and that an animal which lacks the tendency to reinstate auditory stimuli — and thus to imitate sounds — cannot reasonably be expected to talk. This is then compared with the human infant who "exhibits this tendency to a remarkable degree."

The Doctors Gardner tell how experiences by other researchers with chimpanzees were relevant in their own raising of Washoe. Dr. & Mrs. Hayes devised a game with the chimpanzee Viki to "imitate various actions on hearing the command 'do this'."[3] They tell us, "Once established, this was an effective means of training Viki to perform actions that could be visually guided." The Gardners tried this out with Washoe: "Getting Washoe to imitate us was not difficult, for she did so quite spontaneously, but getting her to imitate on command has been another matter altogether."[4]

CHIMPANZEES LEARN BY VISUAL STIMULI

The fact is, chimpanzees learn much more by guided visual stimuli than by the application of auditory methods. Dr. Yerkes, the Hayes, and the Gardners all reported this. With chimpanzees visual stimulus is the primary stimulus for communication purposes.

The picture is somewhat different with young children, for with them visual stimulus is secondary. Children who are born blind utilize semantics (meanings) and syntax (grammar) in an entirely satisfactory manner except for some difficulty with spatial concepts. Professor Eric Lenneberg makes the point that "congenital blindness has no obvious effect on word acquisition, even though there is only a small fraction of words whose referents can be defined tactually."[5]

Another researcher reports that "blind children were found to be four years behind sighted children on tests which required the construction of a series from a set of given lengths."[6]

A CONTRAST BETWEEN BLIND AND DEAF CHILDREN

As we say, blind children utilize both semantics and syntax in a remarkable way, but congenital deafness is another matter unless there is an intensive remediation program. Lenneberg puts it like this: "Congenital deafness has a devastating effect on the vocal facilitation for speech."[7] Despite that devastating effect on primary language abilities, the deaf child can be reached through the visual modality, in the secondary language skill of writing. To quote Lenneberg again concerning the congenitally deaf child, "Presentation of written material enables the child to acquire language through a graphic medium without undue difficulties."[8]

Visual stimuli is tremendously important in the learning process for young children, but that is the general ability of learning as compared with the specific factor of language acquisition. In this presentation we make the point that language is not only a facet of general learning ability, but it also involves a specific ability for which man has an innate capacity.

POINT NO. 5: THE REINSTATEMENT OF AUDITORY STIMULI

We have seen that chimpanzees can be taught to obey and to understand a large number of human speech sounds. That in itself does not suggest ability in speech as such, for it is in the same category as problem-solving and obedience to learned signals, such as a dog would demonstrate.

Nevertheless the chimpanzee shows a great capacity than one would expect of a dog, and in *Apes, Men, and Language* Eugene Linden tells of a chimp named Ally being taught ten words in both English and the American Sign Language. The trainer would say, "Bring me the spoon," and Ally the chimp eventually became proficient and would select the spoon from among a variety of objects.[9]

Another researcher taught Ally the names of several of the same items in American Sign Language, but without having the objects themselves present: "The only stimulus was the instructor's word, 'spoon,' associated with the molding of Ally's hands into a particular gesture."[10] Ally was then asked what a spoon was, and when he was shown it he had to make the correct gesture in the American Sign Language. Linden states, "Among the things Ally had to do to make this appropriate response was to link a gesture, learned from hearing the word, with a stimulus that was visual."[11]

Undoubtedly this ability to transfer from the visual to the auditory mode is a remarkable achievement of learning. Linden himself fully recognizes that there are barriers to human speech as such, for in that same context he suggests that the problem is basically phonological, because "the chimp lacks the necessary mechanisms for generating and controlling particular sounds."

IN APES, "SOUND SIGNALS . . . TRIGGER 'EMOTIONAL' RESPONSES"

However, Linden himself also makes it clear that such an accomplishment is very different from that which is the common experience of every normal human child. Quoting anthropologist Gordon Hewes he suggests that there are severe constraints when chimpanzees are conditioned for such cross-modal transfer of sense data, from the audio-vocal to the visual-haptic channels. He quotes from an article by Hewes entitled, "An Explicit Formulation of the Relationship Between Tool-using, Tool-making, and the Emergence of Language," as follows:

> The inability of apes to acquire even a few quasiarticulate speech sounds does suggest that their cortical auditory centers are somewhat isolated from the centers where visual inputs and voluntary, precise motor outputs are integrated with comparative ease. . . . Sound signals only seem to trigger various holistic "emotional" responses, such as alarm, attention, fear . . . followed by more or less stereotyped behavior patterns such as

flight, attack, protective mothering, submis-
sion, and the like.[12]

Yet another illustration from Linden's book makes the
point clear. One of the chimpanzees that Dr. Roger
Fouts has been training, along the lines that he helped to
train Washoe, is Bruno. He has made very good progress
in the American Sign Language (Ameslan), though at
first he was not at all anxious to mimic the strange
movements he was asked to perform. Fouts was trying to
teach him the sign for the word "hat," involving the
patting of the top of the head, but Bruno would not
cooperate. After a while Fouts got exasperated and
threatened Bruno, whereupon the chimp immediately
started signing, "hat, hat, hat!"[13]

Linden elaborates some of the results with Bruno and
another chimpanzee called Booee and suggests that
"these are just the first teasing glimpses of what Fouts'
next generation of experiments will be about. . . . In the
coming series of experiments Fouts plans to use extraordinary, well-nigh diabolical, reinforcements to encourage the chimps to talk to one another, reinforcements that capitalize on the chimps' appetites and
fears, and, as we shall see, should reveal something not
only about chimp-to-chimp communication using a human language, but also about chimp-to-chimp ethics
when the chimps have to coordinate their efforts to get
some desired reward."[14]

PATIENT JOB KNEW MORE THAN
THE MODERNS

All this is very different from the spontaneous acquisition of language, found in young children wherever
language has been studied around the world.

It is relevant to go back to a quotation that Dr. R. A.
Gardner gives from R. M. Yerkes and B. W. Learned in
Chimpanzee Intelligence And Its Vocal Expression:
"Chim and Panzee would imitate many of my acts, but

never have I heard them imitate a sound and rarely make a sound peculiarly their own in response to mine. . . . Things which are seen tend to be imitated or reproduced. What is heard is not reproduced. Obviously an animal which lacks the tendency to reinstate auditory stimuli — in other words to imitate sounds — cannot reasonably be expected to talk."[15] As Yerkes and Learned go on to say, "The human infant exhibits this tendency to a remarkable degree." All normal humans "reinstate auditory stimuli," not only in imitating but in such activities as recalling something previously said, or remembering a series of audible sounds.

For the chimpanzee, auditory stimulus is secondary, and there is little evidence that chimpanzees reinstate auditory stimulus outside the context of immediacy, or where operant conditioning such as rewards or punishment are involved. As opposed to this, for the human infant auditory stimulus is primary, and the reinstatement of such stimuli is an essential part of the processes involved in human language acquisition.

In this context it is very interesting to notice a twice-made statement in the Book of Job: "Doth not the ear [test] words? and the mouth taste his meat?" (Job 12:11 & 34:3). The picture is fascinating. The tastebuds are at work as we taste our meat, and our ears are essential in the testing, trying, filtering of words. It is strangely modern in its implications. Who told that ancient man Job this modern knowledge about words and language?

Exposure to human speech can be effective when obedience to a command is to be the end result. However, when the intention of such exposure is that there be the acquisition of speech sounds, it is of very little value to expose a chimpanzee to human language.

Because of his experiment of raising the chimpanzee Gua in his own home, Dr. Winthrop Kellogg was in a firsthand position to analyze the possibility of teaching language to a chimpanzee by giving it the same exposure as was given to a normal human child. Dr. Kellogg himself says that at first glance the results were disappointing: "Even in the experimentally controlled environment in which a home-raised chimpanzee (Gua) is given the same linguistic and social advantages of a human baby, the chimp displays little evidence of vocal imitation. Despite its generally high level of imitative behavior, it never copies or reproduces human word sounds."[16]

In earlier chapters we have seen that chimpanzees can understand human words when exposure is sufficient and when reinforcements of various kinds have been applied. Eugene Linden tells us concerning the chimpanzee Lucy being trained by Dr. Roger Fouts, "She seemed to understand spoken English. It was eerie to be talking with Roger about Lucy's mirror or doll and then have her run over and pick it up."[17]

Animals that can reproduce vocal sounds do so at appropriate maturational points. This is as true for members of the monkey family as it is for any other animals. In the article referred to above by Dr. Winthrop N. Kellogg, various chimpanzee noises are referred to, such as the food bark, the "oo oo" cry, and screeching or screaming. Eugene Linden quotes anthropologist Gordon Hewes as describing "the ape's audio-vocal system as basically an 'alarm system' thoroughly tied to the limbic region."[18]

"NOT THE SLIGHTEST EVIDENCE OF TRYING TO REPRODUCE HUMAN VOCALIZATIONS"

Dr. Kellogg outlines the efforts that have been made to teach chimpanzees to adapt to human words. His survey of results by different researchers with the chimpanzees Alpha, Gua, Joni, Fin, Viki, and Washoe is an excellent summary. He tells us that the investigator Kohts, who raised the chimpanzee Joni for approximately two and a half years in the mid 1930's, "noted also that her home-raised chimpanzee displayed not the slightest evidence of trying to reproduce any human vocalizations." In that same context he points out that "no ape has ever been known to go through the long period of babbling and prattling which, in the human baby, seems to be the necessary prerequisite to the subsequent articulation of word sounds."[19]

Dr. Kellogg discusses the measures of success by various experimenters. He quotes W. H. Furness, whose report was published in the proceedings of the American Philosophical Society Vol. 55, No. 281 in 1916, and who finally succeeded to get a young orangutan to say "papa" and "cup:" "Furness found it necessary to place his fingers on the animal's lips and to open and close them in the proper rhythm."[20]

Dr. & Mrs. Keith Hayes used a somewhat similar procedure to teach their chimpanzee Viki to make the sounds of "mama," "papa," "cup," and "up," by manipulating her lips. Even when she had learned to say "mama," "she persisted in putting her own forefinger on her upper lip." Dr. Kellogg tells us that, in Viki's case, these sounds "were learned only with the greatest difficulty. And, even after she could reproduce them, the animal's words were sometimes confused and were used incorrectly."[21] In that same context he makes this illuminating statement: "The most important finding of the Hayeses was perhaps not that their chimp could enunciate a few human sounds. It lay rather in the discovery that these sound patterns were extremely hard for the ape to master, that they never came naturally or easily, and that she had trouble afterward in keeping the patterns straight."

It is relevant to notice that the Kelloggs themselves kept a daily record of their experiments with their ape Gua, and she could obey relatively simple commands such as "Come here," "Close the door," "Go to daddy," "Go to mama," and "Go to Donald," as the case might be. They pointed out that the ape was ahead of the child in the first few months of the experiment but that it was overtaken by the child, "who accelerated at a more rapid rate."[22] He tells us, "Had the comparison continued for

a longer period, all indications are that the human subject would have left the animal far behind in the
comprehension of words."[23]

The results with Gua were very different from what we
find with a child. For the normal human infant to acquire language, exposure to oral human speech is essential, and with speech interaction the child will progressively become efficient in speech communication. That
efficiency is very different from the actions of an animal
taught by reinforcement methods to respond to a stimulus involving human speech.

Dogs will "sit" when so commanded, will carry a
newspaper, go outside, and much more. Similarly the
experiments with chimpanzees make it clear that they
too can learn the appropriate responses to a large number of spoken words. The higher the level of intelligence, the greater is the ability to cope with these
strange sounds from the linguistic world of another species. However, that learning is in a dramatically different category from the spontaneous and sequential
acquisition patterns demonstrated by even nonnormal
human infants.

POINT NO. 6: THE IMPORTANCE OF A SPONTANEOUS ACQUISITION THAT FOLLOWS SEQUENTIAL PATTERN

At this point we see a great contrast between the
teaching of language to the chimpanzee and the spontaneous acquisition of language by the child. With a
chimpanzee there is no spontaneous acquisition, as there
is with the child. The chimpanzee is subjected to various behavioristic patterns such as reward-reinforcement so that a certain word order will be adhered to.
Washoe knows she will get her reward if the signs of the
American Sign Language are produced in the order de-

manded by her human trainers; similarly Sarah knows that she must put her plastic signs in the order demanded of her. Lana knows that unless she presses those keys on her "typewriter" in the correct order there is no hope of getting a reward: if she makes a mistake she presses the "period" key to end the sequence, and then she starts again.

This also is very different from the pattern of spontaneous acquisition found in children across the world. David McNeill comments:

"Children exposed to English and children exposed to Japanese both include abstract grammatical relations in their earliest speech. To these children may be added a third child exposed to Russian. According to Dan I. Slobin (1966) here too, abstract grammatical relations appear in early speech. In spite of radical differences in the conditions of learning, therefore, children are found to do similar things. They do so because of their shared inborn capacity for language."[24]

The pattern of spontaneous acquisition applies also in the realm of phonology. It is widely recognized that children use all the phonemes of all languages in their babbling stage, but for those phonemes to be used as a controlled part of speech it is necessary for the child to have progressive ability in areas involving the tongue, the lips, the palate, the teeth, and so much more. Thus it is that specific cultural sound patterns are uniformly developed, again by spontaneous acquisition as maturational points are reached. As Lenneberg puts it, the development of language is linked with the child's "maturational history."[25]

There is spontaneous acquisition in the realms of semantics, syntax, and morphology, according to maturational patterns. All this adds up to a convincing argument for the distinctiveness of human language. Rudimentary learning patterns are possible for animal exponents of "language," but spontaneous acquisition is denied them.

POINT NO. 7: RELATIONSHIP OF THE PRESENT TO THE PAST AND THE FUTURE

There is another chasm in relation to time. Humans can talk about events, people, and places that are dramatically separated by time and/or space. Basically, animals' communication involves immediacy. They learn from past experiences, and their present behavior is thereby modified, but this is very different from the human ability to relate past, present, and future by words. The "language" of chimpanzees is restricted to the immediate situation. The use of verbs is limited to the present, and immediacy is the order of the day. This is sometimes rather loosely referred to as "displacement," and Eugene Linden suggests, "Displacement actually goes a long way toward explaining what, in fact, are the basic differences between man and chimpanzee."[26]

Linden discusses the case of Ally, the chimpanzee that knew what a spoon was when the word was spoken and also learned to show what it was by a sign. In the learning process it had demonstrated cross-modal transfer, and in a sense this was displacement.

Linden discusses this at some length and includes in his discussion the tool-making ability of the ape: he refers to the well-known case of a chimpanzee peeling the outside from a green twig and then "fishing" into an ant hill and enjoying a tasty respite. Linden makes the point, "All of these suggest that the chimp is capable of demonstrating some of the dividends of displacement, but they leave a teasing question about what constraints limit the chimp's time frame."[27]

That is a major problem. We acknowledge that in some ways (only) the chimpanzee has intelligence approaching that of the early sensorimotor stage of the child, but there is a very great leap between the egocentric activity of peeling a green twig and e.g. working out the engineering techniques to build the Great

Pyramid of Egypt. Linden himself says concerning Roger Fouts' attempts to explore aspects of displacement with the chimpanzees, "This would only be a first step toward establishing how far along the road the chimp is," and, according to Linden, "given the scope of behaviors affected by displacement, it will be only a very small step."[28]

The fact is, all animals are capable of a measure of "displacement," otherwise they would not know when to run for their lives at the sign of danger. Many of the responses of animals result from stimuli which has been impressed upon them in their environment. That stimuli leads to a pattern of behavior over a considerable period of time. Some of the things being called "displacement" with animal training are in that same category: they are associational patterns and are basically in the border area of intelligence and communication.

If it be argued that in language the same concept is there with children, but at a much higher degree, then let it be clearly stated that the difference is dramatic. Humans can talk about their language; they can relate the past from many other cultures as well as the past of their own culture, bringing it into focus to the present; and even at a very early age children can, to a limited extent, "predict" future actions, as when their mother tells them that if they go to sleep they will be taken out tomorrow. Many a child has then gone on to tell the mother what will take place on that promised outing. It is all very different from the limited "displacement" demonstrated by even the brightest chimpanzee.

A SIGNIFICANT CLUE

There is a significant clue in that quotation, highlighting the basic difference between "chimp talk" and human speech. The chimpanzee Ally did *not* learn the word "spoon" ONLY by the auditory clue, but "we helped Ally . . . while moulding his hands into the proper 'spoon' sign."

A point we have already made very strongly is that chimpanzee "language" is primarily manual and involves visual imitation, whereas for human infants auditory stimulus is primary. As Ally learned "spoon" it was NOT only by an auditory clue, but both manual and visual modalities were employed as well. His hands were manipulated (the manual modality), and the sign he was taught bore a visual relationship to the object itself. The learning achievement is remarkable, but it still demonstrates a different ability from what a young child achieves in various forms of "cross-modal" transfer. It was learning a limited amount of language by a combination of stimuli and not by the auditory modality alone. The child learns as he is exposed to and interacts with auditory stimuli — the language by which he is surrounded. The difference is in the inherent capabilities with which the child is born.

ANOTHER WORD ON "CROSS-MODALITY LEARNING"

Before we close this chapter, the matter of cross-modality learning needs to be elaborated. We have included the report earlier in this chapter because we want to be fair in our assessment of the learning achievements of various chimpanzees.

However, there is another aspect that appears to have escaped Eugene Linden's attention. We learn another important fact as we study a report by Dr. Roger Fouts, the man whom Linden interviewed for much of the material in his book. Dr. Fouts took over the training of Washoe when the Gardners concluded their own investigation. Dr. Fouts tells us about the ways he had chimpanzees taught, and he adds various relevant details which are not easily obtained elsewhere. He tells us:

More than 30 student volunteers that I instructed in ASL helped me to teach them 10

signs by physically molding the chimpanzees'
hands into the correct sign form and then giving them a raisin when they made the sign
correctly.[29]

The results were as good as could be expected, though
not at the point of constancy to be found with children
when they use a new word. Fouts tells us,

After teaching the chimpanzees all 10 signs,
we tested them. Some made 90 percent of the
signs correctly, while others only made 26 percent correctly.[30]

Once again we recognize the "poor diction" factor so
ready to obtrude as chimpanzees use human language.

All this is relevant to the "cross-modal" learning
referred to earlier in this chapter. Dr. Fouts gives details of the methods employed:

Initially, we used the vocal word "spoon" to
refer to a spoon. Then we removed the spoon
and taught him to respond to the vocal word
for spoon with its sign, one he did not previously have in his vocabulary. One of us
would tell him, "Sign spoon," or would sign
"Sign" and then say the word "spoon." We
helped Ally learn the new sign by saying
"spoon" while molding his hands into the
proper "spoon" sign. Later, another person
who did not know what sign we had taught
Ally asked him in sign language to give the
signs for several objects that were in the room,
including a spoon. Ally made the correct sign
for spoon. Eventually, he learned 10 signs for
corresponding vocal words and then spontaneously transferred these signs to the objects
that the vocal words represented.[31]

This sort of "double application" is quite unnecessary with the normal child. Indeed, he will often learn a
new word after only one exposure to it. The difference is
in the child's inherent capacity for speech/language, a
capacity radically different from that in the most intelligent chimpanzee.

Footnotes: Chapter 12

1. In *Language In Thinking*, p. 19.
2. *Ibid.*, p. 24.
3. *Ibid.*, p. 24.
4. *Ibid.*, pp. 24, 25.
5. In "A Biological Perspective Of Language" in *Language*, p. 33.
6. In *Language In Thinking*, p. 13.
7. Lenneberg, *A Biological Perspective Of Language*, op. cit., p. 33.
8. *Ibid.*, p. 33.
9. Linden, *op. cit.*, p. 121.
10. *Ibid.*, p. 121.
11. *Ibid.*, p. 122.
12. *Ibid.*, p. 160.
13. *Ibid.*, p. 126.
14. *Ibid.*, p. 134.
15. In *Language In Thinking*, p. 24.
16. *Readings In Animal Behavior*, p. 311.
17. Linden, *op. cit.*, p. 95.
18. *Ibid.*, p. 160.
19. *Readings In Animal Behavior*, p. 311.
20. *Ibid.*, p. 312.
21. *Ibid.*, p. 312.
22. *Ibid.*, p. 312.
23. *Ibid.*, p. 313.
24. "The Creation Of Language," in *Language* (Oldfield & Marshall, Eds.) p. 27.
25. Lenneberg, *Biological Foundations Of Language*, p. 175.
26. Linden, *op. cit.*, p. 158
27. *Ibid.*, p. 166.
28. *Ibid.*, p. 167.
29. Fouts, Roger S., "Talking With Chimpanzees," in *Readings in Psychology* 77/78, Connecticut, Dushken Publishing Group, p. 98.
30. *Ibid.*, p. 98.
31. *Ibid.*, p. 102.

Chapter 13
UNIQUE INFORMATION-PROCESSING

In chapter 11 we talked about "The Unique Human Voice." In that chapter we referred in passing to Eric Lenneberg's emphasis on the brain itself being important, rather than the phenomena associated with articulation.

It is not our purpose to elaborate the functions of the human brain as such, but briefly to consider its information-processing functions as they touch on language communication. We shall see that these are dramatically superior to those of other species.

POINT NO. 8: THE STRATEGIES INVOLVED IN INFORMATION-PROCESSING

Information-processing necessarily utilizes other modalities (channels) as well as the auditory. We have shown that a measure of cross-modal transfer was accomplished with the chimpanzee Ally, and we gave evidence to suggest that this was greatly removed from the spontaneous activity of the normal human child.

Cognitive functions are involved in information-process-ing, and it is widely recognized by primatologists and others that the cognitive functions of the chimpanzee and other members of the ape family operate at a greatly reduced capacity to those shown by humans. It is com-monly stated that the potential of the chimpanzee ap-pears to be analagous to the "sensorimotor" stage of child development: we have seen that this is a term sug-gested by the famous Swiss psychologist Jean Piaget for the stage of development reached by the normal 18 months to two-year-old child. We refer to this again later in this chapter, but for the moment we are thinking especially of information-processing strategies.

"A SET OF CAREFULLY PROGRAMMED LANGUAGE GAMES"

Eugene Linden quotes Roger Brown as writing in *A First Language*, "Processing a sentence which comes to you as simply one from among infinite possibilities of a language seems to be a very different matter from processing that sentence when it arrives as the crowning problem in a pyramid of training which has made one familiar with most of the components involved and put them in a state of readiness."[1] In this context Professor Brown is thinking especially of the chimpanzee Sarah who has put a number of words together in a demanded order so that a certain reward would be available when she had the correct order. Linden quotes Brown as say-ing that these facts are not communication, but "a set of carefully programmed language games."[2]

Ann and David Premack (who raised the chimpanzee Sarah) come remarkably close to a similar conclusion, though expressed rather differently:

The important thing is to shape the lan-guage to fit the information-processing ca-pacity of the chimpanzee. To a large extent teaching language to an animal is simply map-ping out the conceptual structures the animal already possesses. By using a system of naming that suits the chimpanzee we hope to find out more about its conceptual world. Ultimately the benefit of language experiments with animals will be realized in an understanding of intelligence in terms, not of scores on tests, but of the underlying brain mechanism. Only then can cognitive mechanisms for classifying stimuli, for storing and retrieving information, and for problem-solving be studied in a comparative way.[3]

DRAMATICALLY DIFFERENT DEVELOPMENT

As we consider this quotation, a number of facts strike us rather forcibly. There is a recognition of the limitations of the information-processing capacities of the chimpanzee, and it is stated that we should map along the course of the conceptual structures already possessed by the animal, and these processes are included in the concept of "problem solving." In that same context we read that the chimpanzee Sarah "was never put in a situation that might induce such interrogation because for our purpose it was easier to teach Sarah to answer questions."

The information-processing capacities of the human infant are in a dramatically different context — children do not need to be taught to ask questions. As maturational points are reached, questions are asked naturally and spontaneously. Language to the child is not a problem-solving technique but is a remarkably developed ability that goes way beyond anything that the chimpanzee or any other intelligent creation apart from the human species is capable of. The differences are dramatic.

We do not for a moment deny the fact that chimpanzees and other animals have information-processing abilities. It is true that the chimpanzee can label and can even categorize at a simple level. He is capable of perception of a high order, but all this is typical of that sensorimotor intelligence to which Piaget points: In some ways the chimpanzee's achievement is similar to the intelligence of the 18-months-old child. However, the chimpanzee learns by the applicaton of rigorous and unnatural conditions, whereas the child has a specific capacity for language which includes information-processing abilities that are utilized as the appropriate maturational points are reached. By five years of age the child has demonstrated a capacity in this domain of information-processing that is staggering by any standard. The accomplishments of the chimpanzees are minute by comparison.

ANIMAL KNOWLEDGE AND "TRUE CONCEPTS"

Another important point is that children utilize abstract concepts in a way that is quite beyond the capacities of Washoe, Sarah, or any of the other chimpanzees being trained in "language." Dr. Mark P. Cosgrove puts it very well: "Animal signs such as those hand signs of Washoe are not true concepts, since a physical representation is always involved. A true concept, such as the meaning of the words 'political science,' could not be grasped by Washoe through any physical representation. One must not mistake animal knowledge in terms of visual experiences and concrete imagery with true concepts."[4]

The chimpanzees being taught by sign language are capable of combining signs which will express their own wishes or their emotional states. This approaches the threshold of the type of information-processing that takes place with very young children, but a chasm soon separates the accomplishment of the chimpanzees from those of the children.[5]

This author (Wilson) supervised an extensive investigation into the acquisition and development of concepts by 1500 young children in Victorian schools (Australia). By five years of age the majority of the children could answer simple questions dealing with space and time, parts of the body, humans and human activities, classifying and categorizing, the measurement of heat and speed, differences in size and weight, the world and its environment, comparatives and superlatives, numerosity, regular and irregular verb forms, adverbs, noun and pronoun relationships, pluralization, active and passive, past tense, questions of various types, negatives, possessives, and much more.

Those children had no special training, as Washoe had. They were simply normal children with normal exposure to language. They were human: that made all the difference.

These children were not subjected to threats, they were not deprived of their food, nor were they thoroughly spoiled when they were obedient, as it is openly acknowledged that the chimpanzee Lana was spoiled. Theirs was a spontaneous acquisition process resulting from exposure and interchange with other members of the human species. Chimpanzees are subjected to intensive learning processes that are entirely opposed to what they would experience if left to themselves. We again quote from that same survey by Dr. Cosgrove: "In spite of the fact that she mastered all the operations required of her in the laboratory, Sarah's plastic 'words' represent not true concepts but only sensory abstractions. It is true that in sensation an organism will respond to a variety of relationships between objects. However, mere response to relationships does not indicate the presence of abstract thinking of them as relationships."

Notice those words "sensory abstractions." By utilizing methods entirely different from those applicable to children, the chimpanzees have approached the threshold of human communication. They are paddling in a little rock pool while the humans of the same chronological age are effortlessly swimming in the ocean of language which is their natural habitat.

The human is uniquely endowed with superior

abilities in the realms of information-processing. He is also capable of grasping abstract qualities way beyond what any chimpanzee can do. In actual fact, chimpanzees "learn" by dramatically different strategies from those utilized by humans.

Many "behaviorist" psychologists are keen on "behavior modification" strategies to achieve stated objectives in learning. Commendation, rewards, threats, and punishments are all considered legitimate if thereby the set goals can be reached. Such techniques have been openly utilized in the "language learning" of various chimpanzees, but they are quite unnecessary with the normal human infant. This leads to Point No. 9.

POINT NO. 9: SPECIAL AND UNNATURAL CONDITIONS ARE NECESSARY TO TEACH "LANGUAGE" TO CHIMPANZEES

Special conditions, unnatural to the species, are necessary for a chimpanzee to be capable of even minimal language communication with humans. On the other hand, for the human infant language acquisition and learning is entirely natural, and no special conditions are required. It would be unusual indeed for a young child to be threatened with punishment if he did not learn a certain word. As for rewards, before very long many parents would be glad to reward their children for NOT speaking! It would be unnatural for the child NOT to acquire language, without rewards being offered.

That can be contrasted with the way a chimpanzee is "taught" language. Dr. Premack tells us, "When the trainer put several questions on Sarah's board and then walked away, leaving her to answer them, Sarah worked erratically or quit altogether, in somewhat the way a conversation falters when one person ceases to pay attention to the other. Social contact may be Sarah's primary motivation. In any case, she did not work under these circumstances.[6]

In another article Dr. Premack himself makes the point, "In fact, language demands were made of Sarah that would never be made of a child."[7]

Chimpanzees are subjected to various behavior modification techniques in an intensive learning process. This is dramatically different from the natural acquisition of language by any normal child who is exposed to and interacts with language as appropriate maturational points are reached. We shall see that "reward-reinforcements" are very much in evidence in the attempts to teach human language symbols to chimpanzees.

POINT NO. 10: LANGUAGE LEARNING AND REWARD-REINFORCEMENT TECHNIQUES

The November, 1973, *Yerkes Newsletter* shows that Lana then had mastery of 50 different symbols for lexigrams, and that from those symbols she could formulate a variety of requests as she asked her machine for food, liquid, etc. She could gain a great number of things simply by asking. As we saw in chapter 5, her trainers state, "She is the only chimp in a serious research program who with malice aforethought is invited to spoil herself rotten in the interests of advancing science."[8]

This is true of the other chimpanzees as well. For instance, we read of Sarah, "After several of her strategies had failed she paid closer attention to the sentences and began choosing the fruit that gave her the chocolate."[9]

A REWARD FOR A CORRECT RESPONSE

On the same page we further read, "For example, she (Sarah) was given a choice between an apple and a banana, and only when she chose the apple was she given chocolate (which she dearly loved)."

As we have said, these "behavior modification" techniques are common in the training of chimpanzees. Once learning has been successful, sometimes the various communication signs are made without immediate reward, as when Washoe was reading an illustrated magazine and came across a picture of a tiger. She made the sign "cat." According to Eugene Linden the Gardners noted that "when Washoe was not rewarded, she tended more to think aloud rather than talk to the Gardners."[10] Linden suggests this indicates that the infant chimp "is not merely pleasing its masters for rewards or solving puzzles, but is rather turning around and examining its new tool." However, the preliminary use of the tool (in this case language) is very different from the actual process of learning how to handle the instrument or tool in all possible situations and arrangements.

With chimpanzees, behavior modification techniques have been used in ways that simply are not practiced with normal young children. We have already referred to Roger Fouts becoming exasperated and threatening the chimpanzee Bruno, and "Bruno immediately started signing 'hat, hat, hat'."[11] On the very next page of Linden's book there is another clear statement that rewards were very much part of the learning process with the chimpanzees, this time with Booee: "He is a willing learner, in part because of his great love of the raisins that are occasionally used as a reward, but he tests poorly, partly because of the absence of said rewards in testing situations." Bruno learned faster than Booee when rewards were not offered, but Booee was twice as fast as Bruno when raisins were around.

Because of its relevance to this aspect of our study, we again quote Linden's statement relating to a forthcoming series of experiments:

> Fouts plans to use extraordinary, well-nigh diabolical, reinforcements to encourage the chimps to talk to one another, reinforcements that capitalize on the chimps' appetites and fears, and, as we shall see, should reveal something not only about chimp-to-chimp communication using a human language, but also

about chimp-to-chimp ethics when the chimps have to coordinate their efforts to get some desired reward.[12]

What would be the reaction of parents' groups if they read that school teachers were to use "extraordinary, well-nigh diabolical reinforcements" to encourage children to talk to each other — that the teachers were to utilize "reinforcements that capitalize on the [children's] appetites and fears"? There would be many public protests!

CHILDREN ACQUIRE LANGUAGE PROFICIENCY DESPITE "CASUAL AND UNRELIABLE PROCEDURES"

G. A. Miller has this to say:

Finally, we need to make the point that the kind of reinforcement schedule a child is on when he learns language is very different from what we have used in experiments on discrimination learning. No one needs to monitor a child's vocal output continually and to administer "good" and "bad" as rewards and punishments. When a child says something intelligible, his reward is both improbable and indirect. In short, a child learns language by using it, not by a precise schedule of rewards for grammatical vocalizations "in appropriate situations." An experimenter who used such casual and unreliable procedures in a discrimination experiment would teach an animal nothing at all.[13]

That last sentence is an interesting comment. By such "casual and unreliable procedures" an animal would not be taught, and yet the child spontaneously acquires language, with a near-adult skill in syntactics by the time he is five years of age. As Miller goes on to say in that same context:

The child's exposure to language should not be called "teaching." He learns the language, but no one, least of all an average mother, knows how to teach it to him. He learns the language because he is shaped by nature to pay attention to it, to notice and remember and use significant aspects of it.[14]

POINT NO. 11: LEARNING BY DISCRETE LANGUAGE UNITS

"Behavior modification" techniques commonly involve the learning of one new unit at a time. There is a great deal of evidence in the literature concerning the learning of signs by various chimpanzees, indicating that they were learning one sign at a time. They are not expected to extract meaning from a mass of linguistic data, much of which they had never heard before. Nor is there any evidence that they ever put together a sophisticated "linear string." What they do combine is achieved in a way that could have been anticipated as a simple combination of the signs already learned. Careful and patient teaching has meant that the chimpanzees have a large repertoire of such signs and words, and they can put them together in acceptable combinations. Usually this will be for the purpose of gaining rewards, but not always so, as when Washoe made the signs for "listen, dog," when she heard a dog barking in the distance.[15] This is an interesting demonstration of the learning power of a chimpanzee, but it is still very far removed from the natural acquisition of both discrete units and linear strings, which is the regular accomplishment of the normal human infant. It is relevant to notice the emotional content in "listen, dog" for the threat of being given to a dog had been held over Washoe.

A typical example of chimpanzee learning is that given by A. J. & D. Premack: "As a preliminary to learning the class concepts of color, shape, and size, Sarah was taught to identify members of the classes red and yellow, round and square, and large and small."[16] The Premacks elaborate and explain how the concepts of brown and green were taught to Sarah: "For instance, the names 'brown' and 'green' were introduced in the sentences 'Brown color of chocolate' and 'Green color of grape.' The only new words at this point were 'brown' and 'green'."[17]

In another article David Premack tells us further, "As the value of the perceptual class was changed, a corresponding change was made in the language element; and as each new class was mapped, the language requirement was increased."[18]

"EACH NEW WORD . . . THE ONLY UNKNOWN"

Dr. Premack makes the situation very clear in that same article where he tells us, "The basic operation in the training procedure was one-to-one substitution. Each new word or particle was introduced at a marked location as the only unknown in a string of known elements."[19] He further tells us:

One-to-one substitution may be the simplest of all training procedures. When the subject's task is that of completing an admissible string, one-to-one substitution eliminates all of the following possible sources of difficulty:
1. which words to use and how many,
2. where to put the words in the string, and
3. which operations to use — simply addition, or rearrangement and deletion as well.[20]

The situation is quite different with the acquisition of language by a young child. G. A. Miller reminds us, "The meaning of an utterance is not a linear sum of the meanings of the words that comprise it."[21] As the child combines words into sentences, with no set limit (as compared with the usual maximum of eight for the intelligent chimpanzee), all sorts of meanings become possible. The meaning of an individual word is limited. The combinations, as sufficient words are brought together, make possible an infinite number of meanings. The acquisition of both discrete units and linear strings by the human infant is dramatically different in kind from the learning by discrete units of even the most intelligent of the chimpanzee trainees.

POINT NO. 12: INEFFICIENCY WITH "LEARNED" LANGUAGE UNITS

Behavior modification techniques do not always ensure permanent results. It is widely reported that chimpanzees perform accurately about 75 to 80 per cent of the time, with regression to "poor diction" being relatively common. We find such statements as these:

> At first Sarah made many errors, taking the wrong fruit and failing to get her beloved chocolate.[22]

> "Sarah insert apple pail banana dish." Sarah followed the complicated instructions at her usual level of accuracy.[23]

> Sarah performed correctly 75 to 80 percent of the time (her customary level during this phase of the project) and continued to do so when the changes were made.[24]

The same sort of report is made by the Gardners concerning Washoe's learning of language:

> Washoe sometimes fails to use a new sign in an appropriate situation, or uses another, incorrect sign. . . . Again, with new signs, and often with old signs as well, Washoe can lapse into what we refer to as poor "diction."[25]

The Gardner's themselves were very aware there were strict limitations to the potential of their trainee:

On the whole, she has responded quite well to prompting, but there are strict limits to its use with a wild animal — one that is probably quite spoiled, besides. Pressed too hard, Washoe can become completely diverted from her original object; she may ask for something entirely different, run away, go into a tantrum, or even bite her tutor.[26]

"ACQUIRING" RATHER THAN "LEARNING"

The similarities to children in this regard are merely superficial. Instead of having to learn by very intensive application, with great patience on the part of trainers, and by the application of reinforcement and other behavior modification techniques, the child often acquires a word after only one exposure. He acquires rather than learns language, as we have already stated.

One point of great difference as to learning relates to the child's use of irregular words, such as some plurals and past tenses. When the young child says that he "falled over" he is using the same form of the past tense as would be quite correct in "called." When he says "mans" instead of "men" he is using the same plural form that would be correct with "cans." These mistakes indicate that the child is actually rule-oriented, and thus he has a basic approach to language which is unknown to the hard-working chimpanzee.

As Dan Slobin puts it, "A child seeks regularity and is deaf to exceptions. . . . The power of apparent regularities has been noted repeatedly in the children's speech of every language we have studied."[27]

Behavior modification techniques have their advocates, and in this presentation we are not debating the relevance of some of those techniques in various learning situations. However, we do claim very strongly that such techniques are irrelevant for the spontaneous acquisition of language for the normal child who is exposed to and interacts with "everyday" human speech. This is a point of great contrast with chimpanzee language learning.

Footnotes: Chapter 13

1. Linden, *op. cit.*, p. 181.
2. *Ibid.*, p. 181.
3. "Teaching Language To An Ape," *Scientific American*, October, 1972, Vol. 227, No. 4, p. 95.
4. *The Essence Of Human Nature*, Grand Rapids, Zondervan Publishing House, 1977 (in association with Probe Ministries, Texas).
5. See e.g. "The First Sentences Of Child And Chimpanzee," pp. 224-227, in *Psycholinguistics*, Roger Brown, et al, London, MacMillan Ltd. 1970.
6. *Science*, May 21, 1971, Vol. 172, p. 821.
7. *Teaching Language To An Ape*, p. 96.
8. *Yerkes Newsletter*, November, 1973.
9. Premack, A. J. & D., *Teaching Language To An Ape*, p. 96.
10. Linden, *op. cit.*, p. 99.
11. *Ibid.*, p. 126.
12. *Ibid.*, p. 134.
13. Miller, G. A., in "Psychological Aspects Of Linguistics," in *Language*, p. 211.
14. *Ibid.*, p. 211.
15. "Teaching Sign Language To A Chimpanzee," *op. cit.*, p. 38.
16. *Teaching Language To An Ape*, p. 99.
17. *Ibid.*, p. 98.
18. "Language In Chimpanzee?," *Science*, Vol. 172, May 21, 1971, p. 810.
19. *Ibid.*, p. 821.
20. *Ibid.*, p. 821.
21. Miller, G. A., *op. cit.*, p. 207.
22. *Teaching Language To An Ape*, p. 96.
23. *Ibid.*, p. 98.
24. "Language In Chimpanzee?," *op. cit.*, p. 815.
25. *Teaching Sign Language To A Chimpanzee*, p. 25.
26. *Ibid.*, p. 25.
27. *Readings In Human Development*, 1973-1974, p. 130.

Chapter 14
WORDS CAN BE USED IN DIFFERENT WAYS

In this chapter we consider a number of ways that words are used. We find there are basic differences in the ways chimpanzees and children communicate by words, and the differences again highlight the nature of the barriers between the two species.

POINT NO. 13: THERE IS A VERY GREAT QUANTITATIVE DIFFERENCE BETWEEN THE NUMBER OF WORDS LEARNED BY CHIMPANZEES AND THOSE ACQUIRED BY CHILDREN

With chimpanzees, after several years of intensive application, about 100-200 word signals can be utilized: by three years of age the child has a "vocabulary of some 1,000 words."[1] By five years of age the figure has about doubled. In contrast to that, "Sarah has a vocabulary of about 130 terms that she uses with a reliability of between 75 and 80 per cent."[2] We read something similar about Washoe: "Washoe was a good learner. By the time she moved to the Institute for Primate Studies at the

University of Oklahoma, at the age of five, she had not
only learned 132 signs of Ameslan, but had picked up the
rudiments of grammar and syntax."[3] (It is highly
debatable that she had any understanding of "the rudiments of grammar and syntax.")

With continued application Washoe learned more
signs, but her progress is dramatically "retarded" when
compared with a normal five-year-old child.

POINT NO. 14: THE ROLE OF "LABELING"

Another important use of words is to name items in
our environment. Chimpanzees also demonstrate some
ability to label (name) various things, or even people, in
their environment. When Eugene Linden first met
Washoe, she looked at her trainer, Roger Fouts, traced a
question mark in the air and then pointed at Linden.
She was asking who he was.[4]

On another occasion a chimpanzee named Lucy noticed the alligator insignia on Linden's white shirt, pointed to it, and traced a question mark in the air. He gave
back a sign that was meant to imitate the snapping motion of an alligator's jaws. When they came back the
next day Fouts asked Lucy who Linden was, and she
made the sign for "alligator": she had identified the
insignia on Linden's shirts (actually it was on two shirts
on two successive days), and she associated him with the
alligator. In this naming she was able to bring together
qualities of the objects put before her. In a similar way,
she labeled citrus fruits as "smell fruits;" she tasted a
radish, which at first she had called "food," but after she
ate some she called it a "cry hurt food." She referred to
watermelon as a "candy drink."[5]

All this points to a rudimentary form of labeling, and
in fact chimpanzees can also classify objects at a relatively simple level. It is also possible that Lucy invented a sign for "leash," by making the motion of putting a leash on — so that she could be taken out![6] Many

of the errors the chimpanzees made with words were actually within the same category. Thus "Booee erroneously called some foods and fruits 'drinks.' Thelma tended to mix up foods as well."[7]

The fact is, chimpanzees are capable of using word-signs for elementary categorizing and association. These are indications of intelligence, being in that border area in which language and intelligence come together. The two are separate yet intertwined: they cannot be isolated into two totally separate compartments.

"CHILDREN GO ABOUT THE HOUSE . . . NAMING THINGS"

The child's labeling capacity is greatly superior to that of the chimpanzee, and in fact he shows great inquisitiveness as to the actual names of the objects and people in his environment. Early in the acquisition of language he is actively involved in such language, and labeling is an integral part of the young child's life. Roger Brown comments on the fact that at some time in the second six months of life most children say their first intelligible word. He goes on to state, "A few months later most children are saying many words, and some children go about the house all day long naming things (table, doggie, ball, etc.) and actions (play, see, drop, etc.), and an occasional quality (blue, broke, bad, etc.)."[8]

One of the most dramatic stories in the literature about psycholinguistics is that of Helen Keller. When she first realized that her teacher was telling her that the thing falling on her hand was "water," she raced around the garden with her teacher, asking for the names of object after object. A new world had opened to her, and dramatic changes in her personality were soon to follow. When she labeled an object it was thereby identified for her, and it became in a sense an extension of herself.

The fact that children know so many words by the time they are three, as discussed above, indicates that labeling is a dramatic and important part of human activity. At best, chimpanzees have this ability in only a rudimentary form when compared with the highly developed form demonstrated by young children.

POINT NO. 15: WITH CHIMPANZEES, THE NUMBER OF "WORDS" AVAILABLE CAN USUALLY BE CONTROLLED

We have already said that of all the chimpanzee experiments, that with Sarah is the only one that actually uses individual words as such. Washoe's signs are dealing with messages rather than words: when she makes the sign "gimme" a whole message is conveyed, but when Sarah picks up her plastic sign for "apple," she is using a word. Eugene Linden has relevant comments concerning the accomplishments of Sarah: "Dr. Premack's reports leave the impression that what is a medium of communication for her mentors is, from her [Sarah's] point of view, a series of multiple-choice problems."[9]

"SOMEWHAT DIFFERENT FROM NORMAL SPEECH"

Eugene Linden tells us:

Premack admits that both the medium and the methods he used rendered Sarah's language somewhat different from normal speech. Sarah could study the sentences once they were written on the metalized board, something impossible in normal language because of the evanescent nature of the signals. Moreover, Premack notes that "the difficulty of any task can be graded by controlling the number of alternating words available to the subject at any moment in time.[10]

Linden goes on to point out that Washoe must search

through the whole of her vocabulary to formulate her statements, and Sarah at her leisure can select just the right word "from the alternatives her trainers place in front of her."

So far as words are concerned, Sarah is restricted to that prearranged order determined by her trainers. The human child, on the other hand, has a tremendous potential for rearrangement in the sentences that he will create from a very large number of words. Normally the words "available" cannot be controlled by a "trainer." Nor does the child slowly and deliberately choose words over an extended period.

Many other aspects could be dealt with. One is the different meanings possible for words having the same sound, such as "flower" and "flour," "bow" and "bough;" or the same word used either as a noun or a verb, such as "light," or "box;" or words that run together but have quite different meanings according to context, as with "an ice man" and "a nice man," or "a light house-keeper" and "a light-house keeper." The normal child copes adequately and even laughs at ambiguities and absurdities. How would the chimpanzee cope?

POINT NO. 16: THERE IS LITTLE "CREATIVE SKILL" IN THE LANGUAGE LEARNING OF CHIMPANZEES: "THE SUBJECT . . . MERELY USES THE WORDS"

Almost without exception, communication by the "taught" chimpanzees is limited to signals and their meanings as decided by humans. On the other hand, language development in young children demonstrates constant creativity, e.g. as Eric H. Lenneberg puts it, at 24 months of age "all phrases appear to be [their] own creations."[11]

In the main, the chimpanzees were expected to follow the order demanded by their human trainers. Lana would not be given the reward she wanted unless she pressed the keys on her machine in the required order, and Sarah had to put her plastic signs in the order prescribed before she could get her beloved chocolate, or a banana. Dr. Premack tells us, "Notice that, for the test to discriminate between the conjunction and the conditional, Sarah must not impose an order rule of her own."[12] Creativity in language is thus discouraged.

"THE SUBJECT DOES NOT MAKE, BUT MERELY USES, THE WORDS"

Dr. Premack makes the point that the system he has used in teaching a chimpanzee to communicate is different from the human one, in that production does not necessarily lag behind comprehension. He tells us, "The subject does not make, but merely uses, the words; and it can do so from the beginning, without having to undergo elaborate motor learning."[13]

That contains a pointer to the very great difference we are examining, between the language of chimpanzees and children. As Dr. Premack says, "The subject does not make, but merely uses, the words." Children for their part are involved with all sorts of creations and rearrangements of words and phrases. Roger Brown reminds us:

> All children are able to understand and construct sentences they have never heard but which are nevertheless well-formed, well-formed in terms of general rules that are implicit in the sentences the child has heard. Somehow, then, every child processes the speech to which he is exposed so as to induce from it a latent structure. This latent rule structure is so general that a child can spin out its implications all his life long. It is both semantic and

syntactic. The discovery of latent structure is the greatest of the processes involved in language acquisition and the most difficult to understand.[14]

"Latent structure" means that children have an inherent ability to acquire a knowledge of rules and various aspects of the potential of language, and as necessary they spontaneously utilize those rules and bring facets of language together. The chimpanzee can coordinate two or three signs into an association pattern (such as "listen dog"), but this is dramatically different from the creativity displayed as the child utilizes that latent structure which is his heritage as a member of the human race.

We shall elaborate the fact of "rule-orientation" as we proceed. In this chapter we have emphasized basic differences in the use of words as such. The chimpanzee is limited to the words patiently taught him by the members of another species. We saw in Chapter 4 that the child acquires language by simple exposure to and interaction with language as maturational points are reached. He can rearrange words from early childhood in a way that is denied the most intelligent chimpanzee. Human speech in its truest form is species-specific, limited to man who communicates by words brought together into sequences he himself decides.

PUTTING WORDS INTO PROPER ORDER

We have considered a number of similarities and differences in the utilization of words by chimpanzees and humans. In this chapter we consider a number of linguistic principles associated with the meaningful use of words.

POINT NO. 17: THE FIRST WORDS THAT ARE UTILIZED

With chimpanzees, most of the first words in which

proficiency is gained are verbs, whereas with young children the first words are nouns, technically known as "holophrastic expressions." As Roger Brown says, "The nominal . . . is in the early period always a single noun."[15] This is distinctly different from the pattern with chimpanzees, for with them there is a greater emphasis on verbs when only one sign is being utilized in the early stages of teaching.

This becomes clearer when we recognize that the signs themselves are taught to the chimpanzee as though they were verbs. In the list of Washoe's learned word-signs, given by Drs. R. A. & B. T. Gardner in the order of original appearance, we read, "come-gimme;" "more" (explained by the Gardners as meaning, "when asking for continuation or repetition of activities such as swinging or tickling . . . ."); "up" ("wants a lift to reach objects such as grapes . . . ."); "sweet" (among other things, "when asking for candy"); "open" ("at door of house, room, car, refrigerator . . . ."); "tickle" ("for tickling or for chasing games").[16] It could be argued that both the chimpanzee and the child are asking for something: the chimpanzee is asking by action words which often are actually verbs, whereas the child also asks for something, but happens to use nouns in the early stages of language acquisition. There is a point of similarity in that both are making demands, but the difference is that the child's process of using a noun (not a verb) is apparently universal. The noun *is* used as an action word almost invariably, whereas the chimpanzee is learning according to a system imposed on it by adult humans. There is good ground to argue once again that the child's achievement is according to an inherent capacity for the acquisition of language by sequential stages which are similar around the world.

POINT NO. 18: THE FIRST JOINING OF TWO WORDS

Eugene Linden asks the question, "How could

Washoe understand words as symbols without understanding the principles that relate one symbol to another?"[17] In that same context he tells us: "There are significant differences between sign language and spoken language. One is not a mere translation of the other; they have different adaptive functions, and they impose different constraints on communication — gestural language is more telegraphic, less redundant — and, most important, they have different grammars."[18]

Actually even the child does not "understand the principles that relate one symbol to another." As he combines words he correctly follows those linguistic principles, but he does not really "understand" them. Some inherent capacity for language makes correct combining possible, and that "inherent capacity" appears to be limited to humans. The chimpanzee might very well combine words without any understanding of linguistic principles. In the main, he will combine them because of the rewards likely to be made available as a result.

As we relate Linden's two quotations to each other, we find a basic difference between the combinations of words made by Washoe and those made by young children. Linden argues that sign language is basically different from the language acquired by normal human children.

Even as he says, "gestural language is more telegraphic," he is (perhaps unwittingly) highlighting a major difference to be found between the language functions of chimpanzees and children. Chimpanzees minimize their communication (Linden says it is "more telegraphic"), so that one sign will always convey a whole message. On the other hand, as children become more proficient with language they *add* words to the concept they are expressing. Instead of becoming "more telegraphic" they actually elaborate their constructions, as when adjectives are used to make a description more meaningful than with a one-word noun. This is a basic difference in the use of words when we compare a chimpanzee and a child.

WHAT HAPPENS WHEN TWO WORDS ARE JOINED?

The chimpanzee joins two different words, and a second distinct concept is introduced. When the child first uses a second word in a connected two-word utterance, he is elaborating the first concept, very often by using an adjective. As Washoe brings together two signs such as "come, hug,"[19] she is in fact associating two different concepts. When the child moves to the two-word stage, he is following a pattern of rule-orientation. As Roger Brown puts it, "In fact, then, the elaboration of the noun term is accomplished by filling noun positions with just those two-term operations and relations which are noun phrases, and which have been long practiced as independent utterances. It is quite wonderful to find that these first structural complications take just the same form in Finnish as they do in English."[20] This last sentence suggests the possibility that this is a universal facet of language.

THE IMPORTANCE OF STRESS AND PITCH

This type of construction begins at about 18 months of age, and again we quote Roger Brown:

A construction such as *"push car"* is not just two single-word utterances spoken in a certain order. As single word utterances (they are sometimes called holophrases) both *push* and *car* would have primary stresses and terminal intonation contours. When they are two words programmed as a single utterance, the primary stress would fall on *car* and so would the highest level of pitch. *Push* would be subordinated to *car* by a lesser stress and a lower pitch; the unity of the whole would appear in the absence of a terminal contour between words, and the presence of such a contour at the end of the full sequence.[21]

Obviously we do not even think of "stress" and

"pitch" with a chimpanzee. The child uses both quite naturally, for they are part of the pattern as he spontaneously utilizes a whole range of abilities associated with language acquisition.

The joining of two words is part of that pattern, demonstrable in many languages around the world. It is not simply a matter of learning one, and then another, discrete learning unit. That is what happens with chimpanzees. For them, each sign is complete in its meaning, with the following sign equally complete, for it is a new concept. With children, when they first utilize two-word combinations, the new sign will be an extension of the first, elaborating its meaning or clarifying it. The holophrastic one-word "sentence" has given over to a two-word expression that contains the same meaning, but is now somewhat more easily understood by the person who is addressed.

POINT NO. 19: THE RELEVANCE OF WORD ORDER

Because each word is learned as a separate unit, it follows that to a great extent word order with the chimpanzee is haphazard. It is the concept that is important rather than a syntactic pattern, unless there is a reward given for obedience to a particular order. Concerning the chimpanzee he was raising, Dr. Premacks tells us, "there was no conformity between Sarah's errors and those of the trainer. Most of the trainer's errors were systematic — words that he had mislearned, whereas none of Sarah's errors were of this kind."[22]

Once words are put together with human infants, rule-orientation is important. Referring to a particular research program with young chldren, Roger Brown tells us, "violations of order are very uncommon; probably fewer than 100 violations in the thousands of utterances quoted. It is definitely not the case that all possible orders of a combination typically occur; they practically never do."[23]

Professor Dan Slobin tells us further: "Word order is quite inflexible at each of the early stages of syntactic development. One might have predicted that Russian children, being exposed to a great variety of word orders, would first learn the morphological markers for such classes as subject, object, and verb, and combine them in any order. This is, however, hardly the case. Child grammar begins with unmarked forms — generally the noun in what corresponds to the nominative singular, the verb in its adult imperative or infinitive form, and so on. Morphology develops later than syntax, and word order is as inflexible for little Russian children as it is for Americans."[24]

We have said that the child swims in an ocean of language, and yet at a remarkably early age he knows what are the acceptable arrangements of words. Again, this is dramatically different from the reward-reinforcement processes such as those that have been utilized for Sarah with her plastic signs. Even Sarah's grasp of order is linked to egocentric activity and reward techniques. For her, the learning of a certain order is still in the realm of problem-solving, usually to please a human trainer or to gain a desired reward. The Premacks have made it clear that Sarah was taught word order by such reward reinforcement techniques: we saw that "at every stage she was required to observe the proper word sequence."[25]

ORDER IN THE APPEARANCE OF ADVERBS

This is dramatically different from human language learning. The child's word order follows the sequential pattern of acquisition. "Preliminary evidence indicates that the child's adverbs are learned in the order locative, temporal, manner; if true, it might be expected that discourse agreement would proceed in the order where, when, how. In a similar fashion the development of verb restrictions in discourse agreement should follow the development of verb structure."[26]

David McNeill has given some interesting statistics relating to some of the sentences used by the young boy "Adam," in the widely reported study by a Harvard team with three young children. McNeill states:

> One approach is based on the following calculations. With three grammatical classes, there are $(3)^2=9$ different two-word combinations, and $(3)^3=27$ different three-word combinations. If Adam were combining words at random, we should expect to find all (or nearly all) these nine and twenty-seven different combinations. However, they do not all honor the basic grammatical relations. Only four of the two-word combinations directly express one or another grammatical relation, and only eight of the three-word combinations do so. If Adam were attempting to use the basic grammatical relations from the first, he would restrict himself to these admissible patterns. That is exactly what he did. In eight hours of recorded speech, involving some 400 sentences, there were examples of every admissible combination but no examples of inadmissible ones.[27]

Word order develops according to rule orientation which the child could never have learned or invented. This again indicates that there is some innate capacity for language, and even the order of words is important. With those words in proper order the child can produce a virtually infinite number of "correct" constructions.

Footnotes: Chapter 14

1. Lenneberg E. H., "On Explaining Language" in *Science*, Vol. 164, No. 3880, May 9, 1969.
2. *Teaching Language To An Ape, op. cit.*, p. 92.
3. *Newsweek*, March 7, 1977, p. 43.
4. Linden, *op. cit.*, p. 12.
5. *Ibid.*, p. 106.
6. *Ibid.*, p. 109.
7. *Ibid.*, p. 128.
8. *Psycholinguistics*, p. 77.
9. Linden, *op. cit.*, p. 174.
10. *Ibid.*, p. 182.
11. Lenneberg, *Biological Foundations Of Language*, p. 130.
12. In *Language in Chimpanzee?* p. 822.
13. *Ibid.*, p. 822.
14. *Psycholinguistics*, p. 91.
15. *Ibid.*, p. 223.
16. *Teaching Sign Language To A Chimpanzee*, p. 32.
17. Linden, *op. cit.*, p. 73.
18. *Ibid.*, p. 74.
19. *Ibid.*, p. 135.
20. *Psycholinguistics*, p. 223.
21. *Ibid.*, p. 77.
22. In *Language In Chimpanzee?, op. cit.*, p. 822.
23. "The First Sentences Of Child And Chimpanzee," in *Psycholinguistics*, Roger Brown, et al., MacMillan, London, 1970, p. 227.
24. "Grammatical Development In Russian Speaking Children," Dan I. Slobin in *Child Language*, p. 345.
25. *Teaching Language To An Ape, op. cit.*, p. 95.
26. Wick, Miller, and Susan M. Ervin, "The Development Of Grammar In Child Language" in *Child Language*, p. 339.
27. McNeill, D., "The Creation Of Language" in *Language*, p. 25.

Chapter 15
HOW LANGUAGE IS "BROUGHT TOGETHER"

We could have called this chapter, "Morphology, Phonology, Semantics, and Syntax," but it would have seemed too technical for some readers. Actually, those aspects are straightforward after all. They comprise the important ways in which language is "brought together." In our discussion about words, we have considered various aspects of semantics, the meanings associated with words. We shall now briefly consider the other three aspects in turn; and of course, we are still dealing with the ways in which words can be used, as per our chapter heading.

POINT NO. 20: MORPHOLOGICAL ATTACHMENTS

The study of morphology relates to the forms of words. When we add "-ed" to a verb, thereby showing that it is past tense, that "-ed" is a morphological ending. Tenses, plurals, negatives, and so much more are made clear by morphological attachments. These are a normal aspect of sequential language acquisition for the human infant, but they are not part of the repertoire of the chimpanzee.

Eugene Linden discusses the patterning of human language as he compares the achievements of chimpanzees. He tells us, "Instead of a separate signal for each message, human speech uses a finite number of sounds, or phonemes, which can be arranged in a vast number of ways into morphemes."[1]

Morphemes are of two kinds — free and bound. Thus the word we just used ("kinds") is made up of two morphemes: "kind" is a free morpheme, because it can stand on its own and it makes sense. The "s" at the end of "kind" tells us that the word is now a plural. That morphological ending "s" is actually a bound morpheme, for it does not make sense until it is attached to another word.

All this is part of the structure of human languages, with different forms but expressing similar ideas, right around the world, in what are sometimes called their "surface structures" — that is, particular ways in which ideas and thoughts are expressed in a local language or dialect. The human infant has no great difficulty as he discovers morpheme boundaries in the course of his linguistic journey. As Dan Slobin puts it, "From the very beginning of inflections one sees a free use of word stems combined with a huge variety of bound morphemes."[2]

The chimpanzee learns words as messages, complete in themselves. There is no thought of adding other qualities or aspects to those words. On the other hand, as G. A. Miller puts it, "A child does not begin with sounds or words and learn to combine them. Rather, he begins by learning which features are significant, and progressively differentiates his utterances as he learns."[3]

DEVIOUS ROUTES AND PATIENT APPLICATION

By devious routes and patient application of reinforcement techniques, the chimpanzee Sarah was taught to differentiate between objects, and the use of her signs for "same" or "different" are about as close to a morphological attachment as can be demonstrated with chimpanzee "language." By using one or another of those plastic signs Sarah can demonstrate that something is like or unlike something else.

In another context David Premack (who trained the chimpanzee Sarah) refers to the fact that "Man is the only creature with natural language."[4] In that context Premack is actually arguing that the chimpanzee is capable of language, but a language that should be defined in terms other than the forms used by humans as such. The fact is, man's speech and language do possess features that are unique, and that are simply not shared by any other beings on earth. The fineness that is involved with morphology is in that category. Morphological attachments are no problem to humans, because language is natural to man. Morphology is an important aspect of language as it is used by humans. Once again there is a chasm at this point between the normal human child and the most intelligent chimpanzee.

POINT NO. 21: LANGUAGE AS A PHONOLOGICAL SYSTEM

Phonology deals with the sounds of language. The simplest language unit is the allophone, and after that comes the phoneme. The letter "t" is a phoneme, but the "t" in the word "but" is slightly different from the "t" in the word "tub." The "t" in "tub" does not have the plosive content that is in "but." Each "t" is a different allophone, while "t" is the phoneme. However, the phoneme is commonly referred to as the smallest unit of sound. Syllables are made up of a number of phonemes, and so are words. A word may have one or more syllables.

It could be argued that this subject is irrelevant in that we have already seen that the apes do not have the capacity to make the sounds of human language. Nevertheless, the consideration of morphology (discussed above) is incomplete without at least touching on the related area of phonology.

ALL LANGUAGES ARE BASED ON A COMMON PRINCIPLE

Eric H. Lenneberg states, "Phonologically, all languages are based on a common principle of phonematization even though there are phonemic divergences."[5] Chimpanzees do not utilize phonemes, but they make use of word-signs as separate, complete messages. Basically, they are using communication symbols that happen to be human words, but without the fineness and variation possible to humans as they make use of the same word. The nearest to the actual use of words with chimpanzees is in the experiment with Sarah and her plastic signs, and even there the word itself is simply a sign. It is limited to a particular meaning that does not allow for great variation. Dr. Premack himself says in that connection, "There are no phonemes in the language; we have deliberately made the basic unit the word."[6]

This immediately means that there is a dramatic difference between the language taught to Sarah (even with the use of human words) and the language acquired by young children, for phonemic structure is essential to human language as such. Lenneberg tells us further:

The vocalization skills and the behavioral responses to verbal commands that we find in a few species can be shown to bear merely a superficial resemblance to human verbal behavior. In each case it can be demonstrated that their behavior is based on fundamentally different principles from those in humans. The difference is not merely a quantitative one but apparently a qualitative one. . . . No one has demonstrated that a subhuman form can acquire the principles of speech perception in terms of phonemic analysis, of understanding the syntactic structure of a sentence, or of imparting the total semantic domain of any word, be it concrete or abstract.[7]

LIMITATIONS OF "SUBHUMAN FORMS"

Notice especially Lenneberg's statement that, "No one has demonstrated that a subhuman form can acquire the principles of speech perception in terms of phonemic analysis. . . ." This is yet another distinctive of human language as compared with the communication patterns of any animals, including chimpanzees.

The fact is, humans do not only utilize words as separate, complete messages or signals. They can break up words into logical components. They can recombine those parts into new formations, according to the requirements for a particular linguistic expression. Humans utilize principles of phonology and morphology in the arrangement and rearrangement of words and parts of words. This is an ability that is quite beyond the chimpanzees that are being trained in communication by using human language symbols.

POINT NO. 22: HUMAN LANGUAGE IS BASICALLY RULE-ORIENTED: THE IMPORTANCE OF SYNTAX

The material in this section follows what has been given in earlier chapters. We have shown that with chimpanzees a syntactical pattern is largely irrelevant except as a means of gaining a desired reward. We have also seen some of the ways that human language is basically rule-oriented, right from the early stages of advancing from one word to two.

Roger Brown analyzes the structures of the chimpanzee Washoe and suggests that, "We do not yet have evidence that Washoe's sequences are syntactic, because syntax is not just sign combination, but is sign combination employed to express structural meanings."[8] He suggested that Washoe had not discriminated responses in such a way as to indicate that it had distinguished structural meanings, as the child does.[9]

It is true that Sarah and Lana have put word-signs or symbols together in a meaningful connection, but in neither case does this involve syntax as such. A string of symbols put together in proper sequence will lead to a desired reward, and such problem-solving behavior is not really in the same class as the child's use of syntax.

Statements such as the following by David McNeill are commonplace in the literature dealing with psycholinguistics:

> At the age of about one, a normal child, not impaired by hearing loss or speech impediment, will begin to say words. By one-and-a-half or two years, he will begin to form simple two-and-three-word sentences. By four years, he will have mastered very nearly the entire complex and abstract structure of the English language. In slightly more than two years, therefore, children acquire full knowledge of the grammatical system of their native tongue. This stunning intellectual achievement is routinely performed by every preschool child, but what is known about the process underlying it? The process is, as the title of this article implies, one of invention. On the basis of fundamental biological characteristics (of which only slight understanding is presently available), each generation creates language anew.[10]

"UNIVERSALS OF LANGUAGE"

For class lectures to graduate students at Monash University in Victoria, Australia, this author (Wilson) has compiled a list of 50 points which would be widely recognized among psycholinguists as "universals" of language.[11] Most of them could be included under the subheading of this section relating to human language being rule-oriented. The points there outlined touch on phonology, syntax, semantics, and morphology. Sequential patterns, maturational stages, the recognition of glaring ambiguities, the "embedding" of new concepts (as with "That big dog lives with the old man" instead of "The dog that is big lives with the man who is old") are all part of the child's rule-orientation patterns. Rules are spontaneously utilized by the child as he extracts his personal constructions from the language by which he is surrounded. The difference in achievement by the chimpanzee is overwhelmingly great.

How, then, is language "brought together?" Behaviorist psychology does not give a satisfying answer. Neither do the rationalists such as Noam Chomsky, for they do not explain the "how?" Only the theistic creationist can do that. However, Chomsky and his school have at least described a great deal about the processes taking place. They have described, rather than explained, but their description makes a lot of sense as to essentials.

We recognize that "deep structures" go back to both semantics (base meanings) and syntax (rule-oriented speech). This simply highlights the marvel of the functions involved. The reality is that humans are uniquely endowed with a species-specific capacity for the acquisition and learning of language as it is used by human beings around the world. We consider that in our next section.

POINT NO. 23: AN INHERENT CAPACITY FOR LANGUAGE ACQUISITION IN HUMANS

Psycholinguists such as Noam Chomsky and Eric H. Lenneberg argue that humans appear to have a unique,

species-specific, and species-universal language acquisition system. This system is innate and is triggered by
the language environment surrounding the child as
appropriate maturational points are reached. Where this
"system" is adversely affected, as by brain surgery, language learning (but not simple acquisition after
puberty) can still take place.

THE IMPORTANCE OF
ERIC H. LENNEBERG

Language is basically associated with the left
hemisphere of the brain, and Professor Eric H. Lenneberg's work is important in this area. The following
notes by this author (Wilson) have been used with
postgraduate classes at Monash University in Victoria,
Australia. They are summarized from Dr. Lenneberg's
classic *Biological Foundations of Language*, and from an
article, *On Explaining Language*, Lenneberg published
in *Science* of May 9, 1969.

The points made are of great interest in their own
right, and they are also highly relevant to the present
discussion as to the role of a "Language Acquisition
Device" in humans. ("System" would be better —
"Device" has unfortunate overtones suggesting something like a little black box in the brain.)

A number of other aspects could be listed, but the
following 15 points are especially relevant:

1. If a child's brain is traumatized after the onset of
   language, but before the age of four, his language
   can be quickly reestablished.
2. Such a child's language development will be
   similar to that of the two-year-old but at a faster
   rate.
3. Lesions before teens have an excellent chance of
   recovery as regards language.
4. After midteens it becomes increasingly difficult for
   language to be attained.
5. The acquisition of a foreign language after late

teens is much more difficult than before midteens — accent problems and interference of the mother tongue become increasingly real problems: "Something happens in the brain during early teens that changes the propensity for language acquisition."[12] (However, a matrix for the first language has been established, and linguistic principles can be transferred in further language learning.)

6. Early teens is apparently a critical period that "coincides with the time at which the human brain attains its final state of maturity in terms of structure, function, and bio-chemistry."[13]

7. "Apparently the maturation of the brain marks the end of regulation and locks certain functions into place.[14]

8. Every stage of maturation is unstable and prone to change which can take place by a triggering from the environment.

9. "With continued exposure to these patterns as they occur in a given language, mechanisms develop that allow him to process the patterns, and in most instances to reproduce them. . . ."[15]

10. The sequence of acceptance, synthesis, and the state of new acceptance can be demonstrated for both semantics and phonology.

11. There is a correlation between anatomical, physiological, cognitive, and language development.

12. Man's "entire cognitive function, of which his capacity for language is an integral part, is species-specific."[16] Comparisons of language with animal communications are invalid, with similarities resting on superficial intuitions.

13. Mongoloid children develop physically beyond teens, but progressive language development does not. This is an important pointer to the fact that language *acquisition* gives over to *learning* at teens.[17]

14. "Raw material" for language is the language

environment surrounding the child and "seems to function like a releaser for the developmental language synthesizing process."[18]

15. Lenneberg has lent support to Chomsky's hypothesis of an inherent "Language Acquisition Device." Others besides Lenneberg support Chomsky in arguing for an inherent language ability in humans. Thus, in an article entitled "The Creation of Language By Children," David McNeill discusses the fact of language acquisition across various languages and states, "It appears that children do identical things in the face of radically different conditions of learning. The proposal that linguistic theory represents children's inborn capacity for language accordingly gains empirical support."[19]

"EMPIRICIST THEORIES OF LEARNING ARE QUITE INADEQUATE."

Noam Chomsky argues, "The study of language, it seems to me, offers strong empirical evidence that empiricist theories of learning are quite inadequate. Serious efforts have been made in recent years to develop principles of induction, generalization, and data analysis that could account for knowledge of a language. These efforts have been a total failure. The methods and principles fail, not for any superficial reason such as lack of time or data. They fail because they are intrinsically incapable of giving rise to the system of rules that underlies the normal use of language."[20]

Chomsky's statement that efforts to account for the knowledge of language "have been a total failure" is very strong. Nevertheless it is indeed true that the typical behaviorist hypotheses do not satisfactorily account for the emergence of language in young children. Ultimately, we are forced to seek explanations elsewhere, and there seems to be no alternative to the acceptance of the fact that humans are uniquely endowed with an innate capacity for language as it is used by humans around the world.

Chomsky also states, "What evidence is now available supports the view that all human languages share deep-seated properties of organization and structure. These properties — these linguistic universals — can be plausibly assumed to be an innate mental endowment rather than the result of learning. If this is true, then the study of language sheds light on certain long-standing issues in the theory of knowledge. Once again I see little reason to doubt that what is true of language is true of other forms of human knowledge as well."[21]

It is beyond our present purpose to investigate philosophical issues involved in theories of language acquisition, but we notice in passing that an interesting possibility is hinted at in the above quotation. If the capacity for language acquisition is innate in humans, what about what Chomsky calls "other forms of human knowledge as well"?

"ATTRIBUTE THIS TO EVOLUTION . . . NO SUBSTANCE TO THIS ASSERTION"

He goes on to suggest that, "The process by which the human mind has achieved its present state of complexity and its particular form of innate organization are a complete mystery. . . ."[22] He then writes, "It is perfectly safe to attribute this to evolution, so long as we bear in mind that there is no substance to this assertion — it amounts to nothing more than the belief that there is surely some naturalistic explanation for these phenomena."[23]

Let it be clearly stated that Noam Chomsky would not endorse many of the views of this present writer, but yet it is a remarkable fact that the rationalist argument he puts forward comes very close to the creationist point of view (which this author holds).

Ultimately, the debate between the behaviorists and

the rationalists, of whom Chomsky is a leader, is on the question of what is innate. Both necessarily accept the concept of innateness, but behaviorists insist that only general learning strategies are innate. They see one basic learning process, innate in all living beings whether they be pigeons, chimpanzees, or humans. The rationalist school, of which Chomsky is a leader, argues that language is a specific innate ability in humans. The ability for language acquisition is a separate quality, according to this argument, separate from the general learning potential of an individual that varies according to his intelligence. Many rationalists would argue (with considerable evidence) that language and intelligence are related yet separate: it is a fact that the person with high intelligence might have low language abilities and vice versa.

Chimpanzees have intelligence and general learning abilities, limited though they are (by comparison with those of human infants) when attempting activities especially suited for humans as such. Humans on the other hand are uniquely endowed with a specific language ability which means that even the "below average" child will acquire language progressively until about the onset of puberty. From that time onward language must be learned in the same way as any other human skill is learned — by diligent application, and not by the relatively easy process of exposure and interaction, as is the case with the acquisition of language by normal human children.

We have suggested that the rationalist argument comes close to the creationist point of view. We have quoted Chomsky as to the "complete mystery" of how the human mind "achieved its present state of complexity and its particular form of innate organization." We have noted his statement that, "It is perfectly safe to attribute this to evolution, so long as we bear in mind that there is no substance to this assertion — it amounts to nothing more than the belief that there is surely some naturalistic explanation for these phenomena."

There is an alternative, unacceptable to most academics, partly because they will not open their minds to the possibility. We refer to the Christian concept that man was directly created by Almighty God, created to be the friend of God, able to commune with Him by speech/language.

Footnotes: Chapter 15

1. Linden, *op. cit.*, p. 138.
2. Slobin, Dan I., "Grammatical Development In Russian Speaking Children" in *Child Language*, p. 346.
3. Miller, G. A., *Psychological Aspects Of Linguistics* in *Language* p. 209.
4. *Language In Chimpanzee?, op. cit.*, p. 822.
5. "A Biological Perspective Of Language" in *Language*, p. 34.
6. *Language In Chimpanzee?, op. cit.*, p. 809.
7. Lenneberg, *A Biological Perspective Of Language, op. cit.*, p. 33.
8. *Psycholinguistics*, p. 227.
9. *Ibid.*, p. 226.
10. McNeill, D., "The Creation Of Language" in *Language*, p. 21.
11. An outstanding text is *Universals of Language*, Joseph H. Greenberg (ed.), M.I.T. Press, Cambridge, Mass., 1963 (4th printing, 1973).
12. "On Explaining Language" in *Science*, May 9, 1969, Vol. 164, No. 3880, p. 639.
13. *Ibid.*, p. 639.
14. *Ibid.*, p. 639.
15. *Ibid.*, p. 641.
16. *Ibid.*, p. 642.
17. *Biological Foundations Of Language*, pp. 154-155, 309 ff.
18. *Ibid.*, p. 375.
19. In *Child Language*, (Ed Bar-Adon), p. 356.
20. Noam Chomsky article "Language And The Mind" in "Child Language," pp. 432-433, as reprinted from *Psychology Today*, February, 1965.
21. *Ibid.*, p. 432.
22. *Ibid.*, p. 433.
23. *Ibid.*, p. 435.

CONCLUSION

The various experiments with chimpanzees have been extensions of problem-solving techniques, utilizing methods that are not primary for the spontaneous and sequential acquisition of speech/language by all normal babies around the world.

Other created beings communicate in various ways, utilizing various senses and body organs, but only man is able to communicate by means of true speech/language.

Such speech is still "species-specific," confined to man who is made in the image of God. Because of this uniqueness, man is able to communicate with his fellow-man and also to worship his God.

BIBLIOGRAPHY

Adams, Parveen, *LANGUAGE IN THINKING*, (Penguin, 1973).

Adcock, C. J., *FUNDAMENTALS OF PSYCHOLOGY*, (Penguin Books, 1959).

Al-Issa, I., "The Development of Word Definition in Children" in *The Journal of Genetic Psychology*, Vol. 14, 1969.

Allen, J. P. V. & Van Buren, Paul, *Chomsky: Selected Readings*, (Oxford University Press, New York, 1972).

Altmann, Stuart A., "Auditory Communication Among Vervet Monkeys" in *Social Communication Among Primates*, (University of Chicago Press, 1961).

Altmann, Stuart A. (Ed.), *Social Communication Among Primates*, (University of Chicago Press, Chicago, 1967).

Aronson, L., et al, *Development and Evolution of Behavior*, (Freeman, New York & San Francisco, 1970). Elaborated by R. A. Hinde in *Non-Verbal Communication*.

Bar-Adon, A. (Ed.), and W. F. Leopold, *Child Language*. Elaborated by Dan Slobin. (Prentice-Hall, Inc., Englewood Cliffs, NJ, 1971).

Bar-Adon, Aaron, *Primary Syntactic Structures in Hebrew Child Language*.

Borger, Robert & Seaborne, A. E. M., *The Psychology of Learning*, (Penguin Books, 1966).

Bower, T. G. R., "The Visual World of Infants," Dec., 1966, in *Pre-natal Development and Capacity of the Newborn*.

Bracken, Harry M., "Chomsky's Variations on a Theme by Descartes," *Journal of the History of Philosophy*, VIII, No. 2 (April, 1970).

Britain, James, *Language & Learning*, (Penguin Books, Harmondsworth, Middlesex, England, 1972).

Bronowski, J. & Ursula Bellugi, "Language, Name, & Concept," in *Science*, Vol. 168, May 8, 1970.

Brown, Roger, et al., *Psycholinguistics*, paper entitled "The First Sentences of Child & Chimpanzee." (The Free Press, New York [Collier, Mackvilla, Ltd.], 1972).

Brown, Roger, & Ursula Bellugi-Klima "Three Processes in the Child's Acquisition of Syntax," in *Child Language*.

Brown, Roger, Courtney Cazden & Ursula Bellugi-Klima, *The Child's Grammar from One to Three*.

Cazden, Courtney B. *Child Language & Education*, (Holt, Rinehart & Winston, New York, 1972).

Cazden, Courtney B., *Child Language & Acquisition*, (Holt, Rinehart & Winston, New York, 1972).

Chomsky, Carol, "Language Development After Age Six" in *Readings in Child Behavior & Development*, (Celia Stendler Lavatelli & Faith Stendler, Harcourt, Brace, Jovanovich, Inc., New York, etc.). Reprinted from *The Acquisition of Syntax in Children from Five to Ten*, by Carol Chomsky, M.I.T. Press, Cambridge, Mass., 1969.

Chomsky, Noam, "The Formal Nature of Language," being Appendix A in Lenneberg, *The Biological Foundations of Language*.

Chomsky, Noam, *Language & Mind*.

Chomsky, Noam, *Language & The Structure of Mind in Language in Thinking*, (ed. Parveen Adams); also "Recent Contributions to the Theory of Innate Ideas," first published in *Boston Studies in the Philosophy of Science*, Vol. III, (Humanities Press, New York, 1968). Reprinted in "The Philosophy of Language," (ed.) J. R. Searle, *Oxford Readings in Philosophy*, (Oxford University Press, 1971).

Chomsky, Noam, "Linguistic Contributions to the Study of Mind: Future," in *Language in Thinking.*

Chomsky, Noam, *A Review of B. F. Skinner's Verbal Behavior in Language,* 1959. Reprinted in Fodor & Katz (eds.) 1964.

Clark, Matt & Shapiro, Dan, "Almost Human," *Newsweek,* March 7, 1977.

Cosgrove, Mark P., *The Essence of Human Nature,* (Probe Ministries International, Dallas, Texas, 1976).

Davey, John, "Can Chimps be Taught to Speak?" (*Observer Review,* Aug. 10, 1969).

Deese, James, *Psycholinguistics,* (Allyn & Bacon, Inc., Boston, 1971).

deHaan, Bierens, J. A., "Animal Language & Its Relation to That of Man," (*Biological Review,* 1929). *Philosophical Works of Descartes,* Vol. 1 trans. by Elizabeth S. Haldane & G. R. T. Ross (Cambridge University Press, London, 1969). Meditation VI.

Duffy, Gerald G., *Teaching Linguistics.*

Fabricus, Eric, "Crucial Periods in the Development of the Following Response in young Nidifugous Birds," in *Readings in Animal Behavior.*

Firth, H. G., *Influence of Language on the Development of Concept Formation in Deaf Children,* and H. Sinclair-de-Zwart, *Developmental Psycholinguistics.*

Foss, Brian, *New Perspectives in Child Development,* (Penguin Books, 1974).

Frisch, Karl V., "Honey Bees: Do They Use Direction & Distance Information Provided by Their Dances," in *Readings in Animal Behavior.*

Gardner, R. A. & B. T., "Teaching Sign Language to a Chimpanzee," in *Language in Thinking,* (ed.) Parveen Adams, Table I, *Science,* 1969, Vol. 165, No. 3894.

Gilbert, B., *How Animals Communicate*. (Angus & Robertson, Ltd., London, 1966).

Gleason, Jr., H. A., *Linguistics & English Grammar*, (Holt, Rinehart & Winston, Inc., New York, 1965).

Gleason, Jean Berko, *The Child's Learning of English Morphology*.

Greenberg, Joseph H., *Universals of Language* (ed.) (M.I.T. Press, Cambridge, MA, 1963, 4th printing, 1973).

Greene, Judith, *Psycholinguistics: Chomsky & Psychology*, (Penguin Books, Baltimore, 1972).

Hannah, Elaine P., *Applied Linguistic Analysis:Stages One, Two*.

Hayes, C., *The Ape in Our House*, (Harper & Rowe, New York, 1951).

Hebb, Lambert & Tucker, *Psychology Today*, April, 1973 "A D.M.Z. in the Language War."

Hinde, Robert A. *Animal Behavior: A Synthesis of Ethology & Comparative Psychology* (McGraw-Hill Book Co., New York, 1970).

Hinde, Robert A. (ed.) *Non-Verbal Communication*. (Cambridge University Press, London, 1972).

Hockett, Charles F. "Logical Considerations in the Study of Animal Communication," in *Animal Sounds & Communications* (ed.) W. E. Lanyon & W. N. Tavolga, (American Institute of Biological Sciences, Washington, D.C., 1960).

Hockett, Charles F., *The Universals of Language*, (ed.).

Hughes, Gordon, "Primate Communication & The Gestural Origin of Language," in *Current Anthropology*, Vol. 14, Nos. 1-2 (University of Chicago Press, 1973).

Hurlock, Elizabeth B., *Child Development*, (McGraw-Hill Book Co., New York, 1964).

Jakobson, Roman & Morris, *Phonemic Patterning*.

John, Vera P. & Sarah Noskovitz, "Language Acquisition & Development in Early Childhood," in *Linguistics in School Programs*, the 1970 Yearbook of NSSE.

Keller, Helen, Supplement to *The Story of My Life,* (Grosset & Dunlap, New York, 1905).

Kellogg, Winthrop N., *Communication & Language in the Home-Raised Chimpanzee.*

Klima, Edward W. & Ursula Bellugi-Klima, *Syntactic Regularities in the Speech of Children.*

Klopfer, P. H. & M. S., "Material 'Imprinting' in Goats: Fostering of Alien Young," in *Readings in Animal Behavior.*

Krech, David, *Psychoneurobiochemeducation.*

Lambert, W. E. & T. Richard Tucker, "A D.M.Z. in the Language War," *Psychology Today,* April, 1973.

Langer, Susanne, *Philosophy in a New Key,* Mentor Books, New York, New American Library, 1952.

Lenneberg, E. H., "A Biological Perspective of Language" in *Language* (eds.) Oldfield & Marshall.

Lanneberg, E. H. (ed.) "A Biological Perspective of Language," in *New Directions in the Study of Language,* M.I.T. Press, Boston, 1964.

Lenneberg, Eric H., "On Explaining Language," in *Science,* May 9, 1969.

Lenneberg, Eric H., "Language in the Light of Evolution" in *Animal Communications* (Thomas A. Sebeok (ed.) Indiana University Press, Bloomington, 1973).

Langacker, Ronald H., *Language & Its Structure,* (Harcourt, Brace & Jovanovich, N.Y., 1973).

Lee, Laura, "Comparing Syntactic Development" in *Journal of Speech and Hearing Disorders,* XXXI, 4 (1966).

Lee, Laura L., *Developmental Sentence Analysis,* Northwestern University Press, Evanston, 1974.

Leopold, Werner F., *The Study of Child Language & Infant Bilingualism,* Jakobson, Roman, *The Sound Laws of Child Language & Their Place in General Phonology.*

Lester, Mark (ed.) *Readings in Applied Transformational Grammar,* (Holt, Rinehart & Winston, New York, etc., 1973).

Lieberman's article, "Primate Vocalizations & Human Linguistic Ability," reprinted from the *Journal of the Acoustical Society of America*, Vol. 44, No. 6, Dec., 1968.

Lilly, John C. & Alice M. Miller, "Vocal Exchanges Between Dolphins," in *Animal Social Psychology*, Robert B. Zajouc (ed.), John Wiley & Sons, Inc., New York, 1969.

Linden, Eugene, *Apes, Men & Language*, Penguin, New York, etc., 1974.

Logan, Lillian M. & G. Virgil, *A Dynamic Approach to Language Aids*, McGraw-Hill, New York.

Lorenz, Konrad, *Studies in Animal & Human Behavior*, Vol. 1, Methuen & Co., Ltd. N.Y., 1970.

Luria, A., *The Working Brain*, Penguin, England, etc., 1973.

Lyons, John, in Hinde op. cit., See *Animal Communication*, Indiana University Press Bloomington, 1968. (ed.) T. A. Sebeok.

Lyons, John, "Human Language" in *Non-Verbal Communication* (ed.) Robert A. Hinde.

McGill, Thomas E., *Readings in Animal Behavior.*, Holt, Rinehart & Winston, Inc. New York, 1973 2d ed.

McNeill, David, "The Creation of Language by Children," in *Child Language*. (ed.) Oldfield & Marshall.

Menyuk, Paula, *The Acquisition & Development of Language*, Prentice Hall, NJ, 1971.

Mistler-Lachman, Janet L. & Roy Lachman, "Language in Man, Monkeys, & Machines," in *Science*, Vol. 185, Sep. 6, 1974.

Oldfield, R. C. & J. C. Marshall (eds.) *Language*, Penguin Education, England, 1973.

Paton, Andrew, Postgraduate student at Monash University, Victoria, Australia (1977), utilizing the Wilson Concept Development Scheme.

Pavlov, I. P. *Psychology, Conditioned Reflexes,* translated G. V. Anrep, London, Oxford University Press, 1927.

Pengelly, Eric T. & Sally J. Asmundson, "Annual Biological Clocks" in *Readings from Scientific American — Animal Behavior* introduced by Thomas Eisner Cornell & Edward O. Wilson, W. H. Freeman & Co., San Francisco, 1975.

Pflaum, Susanna Whitney, *The Development of Language & Reading in the Young Child,* Charles E. Merrill Publishing Co., Columbus, Ohio, 1974.

Pincas, Anita, "Transformational Generative & The E.F.L. Teacher," in *Linguistics in the Elementary School Classroom* (ed.) Paul S. Anderson.

Premack, A. J. & David Premack, "Teaching Language to an Ape" in *Scientific American,* October, 1972, Vol. 227 (4).

Premack, David, "Language in Chimpanzee?" Reprinted from *Science,* Vol. 172 of May 21, 1971.

Putnam, Hilary, *Symposium on Innate Ideas* (ed. J. R. Searle), Oxford University Press, London, 1971.

Rosenberg, Jay F. & Charles Travis (eds.) *Readings in the Philosophy of Language,* Prentice Hall, Inc., Englewood Cliffs, NJ, 1971.

Rumbaugh, Duane M. (ed.) *Chimpanzee — Gibbon & Siamang,* Vol. 3: *Natural History, Social Behavior, Reproduction, Vocalizations, Comprehension,* published by S. Karger, NY, 1974.

Rumbaugh, Duane M., *Natural History, Social Behavior, Reproduction, Vocalization, Comprehension.,* S. Karger, NY, 1974.

Searle, J. R. (ed.) *Symposium on Innate Ideas — The Philosophy of Language,* Oxford University Press, London, 1971.

Sebeok, T. A. & A. Ramsay (eds.) "Semiotics & Ethology," in *Approaches to Animal Communication,* 1968, Indiana University Press, Bloomington.

Selsam, Millicent E., *The Language of Animals*, Kingswood, Ltd., Tadworth, Surrey, England 1962. World's Work, Ltd.

Schaller, G. B., *The Mountain Gorilla, Ecology, & Behavior*, University of Chicago Press, Chicago, 1963.

Scott, John Paul, *Animal Behavior*, Univ. of Chicago Press, Chicago, 1972.

Singh, Sheo Dan, "Urban Monkeys," in *Scientific American*, 221/1, No. 7, 1969.

Slobin, Dan, "On the Nature of Talk to Children," published in *Foundations of Language Development*, E. H. & E. L. Lenneberg (eds.) Academic Press, NY, 1975.

Slobin, Dan, article on the noun and the acquisition of articles in Paula Menuk, *The Acquisition & Development of Language*.

Slobin, Dan I., *Psycholinguistics*, Scott-Foresman, New York, 1971.

Slobin, Dan I., "They Learn the Same Way All Around the World," in *Readings in Human Development*, 1973-74. (Reprinted from *Psychology Today*, July, 1972).

Struhsaker, Thomas T., "Auditory Communication Among Vervet Monkeys," in *Communication Among Primates* (ed.) Stuart A. Altmann, University of Chicago Press, Chicago, 1967.

Templin, M., *Certain Language Skills in Children: Their Development & Inter-Relationships,* University of Minnesota Press, Minneapolis, 1957.

Thomson, Robert, *The Psychology of Thinking.* (Penguin Books, 1959).

Wilson, Clifford, *The Measurement of Psycholinguistic Quotient in Young Children*, University of SC, Columbia, SC.

Wilson, Clifford, *Language Abilities Guide*, 1975. Worth Publishing Co., Melbourne, 1976.

Wilson, Clifford, *Language: Is it Acquired or Learned? A study based on the Wilson Language Abilities Guide*, 1976. Monash University, Victoria, Australia.

Wilson, Clifford, *That Incredible Book: The Bible*, Word of Truth Productions, Box 288, Ballston Spa, NY 12020.
Exploring Bible Backgrounds
Exploring the Old Testament
New Light on the Gospels
New Light on New Testament Letters
Jesus the Teacher
UFO's and Their Mission Impossible,
Crash Goes the Exorcist

Published by Creation-Life Publishers, San Diego, CA
Crash Go the Chariots
Gods in Chariots & Other Fantasies
East Meets West in The Occult Explosion
Ebla Tablets: Secrets of a Forgotten City
The Passover Plot Exposed

Wood, Barbara S., *Children & Communication*, Prentice-Hall, NJ, 1976.

Yerkes Newsletters, Vol. 9-22, Sept. 1972, Vol. 10-2, Nov., 1973, & Vol. 11-1, May, 1974.

Yerkes Newsletters, Vol. 9-2, Sept., 1972.

Zajanc, Robert B., *Animal Social Psychology*, John Wiley & Sons, Inc., NY, 1969.

Other Books of Interest
from
MASTER BOOKS
P. O. Box 15666
San Diego, California 92115

CLOSE ENCOUNTERS: A Better Explanation
Clifford Wilson, Ph.D. and John Weldon

Are UFO's really of extraterrestrial origin? Are they Biblical? Is the movie "Close Encounters" science fiction? Are aliens trying to bring us messages from other galaxies? Have they really abducted earth people? Read an alternate point of view on the subject of UFO's and their occupants. Tells about actually documented "encounters." No. 034, Paper $2.95

Approaching the DECADE OF SHOCK!
Clifford Wilson, Ph.D. and John Weldon

1980's!! The forthcoming decade of shock. This has been predicted by Bible scholars, psychologists, astrologers . . . and can be foretold by anyone who reads or listens to the news daily. Read about fulfilled prophecy, the scriptural and spiritual impact of current events, and what the 1980's will bring. No. 004, Cloth $5.95

War of the Chariots
Clifford Wilson, Ph.D.

Resulting from a debate with Erich vonDäniken, this book refutes vonDäniken's latest absurd accusations that Jesus was an astronaut from outer space. Dr. Wilson is the author of the million-copy best seller, *Crash Go The Chariots*. No. 183, Paper $2.95

The Troubled Waters of Evolution
Henry M. Morris, Ph.D.

The most complete account, from a Biblical and creationist point of view, of the long history of evolutionary thought, with a non-technical study of the evidence for creation, especially the second law of thermodynamics. No. 170, Paper $2.95

Crash Go The Chariots
Clifford Wilson, Ph.D.
Million-copy best seller reply to the farfetched imaginings
of vonDäniken. Enlarged and updated from the original
edition. Provides sensible answers to the absurd claims of
Chariots of the Gods. Includes a segment on the Bermuda
Triangle.   No. 035, Paper $1.95

East Meets West in THE OCCULT EXPLOSION
Clifford Wilson, Ph.D.
Here is real help in these confusing times in dealing with
spiritism, witchcraft, transcendental meditation, yoga,
Hare Krishna, and other Satanic cults.
                                                   No. 113, Paper $1.95

Ebla Tablets: SECRETS of a Forgotten City
Clifford Wilson, Ph.D.
The greatest find since the Dead Sea Scrolls. An abun-
dance of clay tablets in a buried city in the Tell Mardikh
contain such fascinating information as trade documents
referring to the Garden of Eden (Dilmun) as an actual
geographical location, a creation tablet remarkably similar to the Hebrew record—*in writing* more than 2000
years B.C., and much more.   No. 052, Paper $1.95

The Passover Plot—EXPOSED!
Clifford Wilson, Ph.D.
An expert challenge to the confused (but widely accepted)
imaginings of the book and movie, *The Passover Plot.*
Wilson substantiates that Christians have not been de-
ceived by a *hoax,* but that the death and resurrection of
Jesus, rather than a *plot,* were part of the greatest *plan* of
all ages.   No. 125, Paper $1.95

Crash Goes the Exorcist
Clifford Wilson, Ph.D.
An incisive analysis of both the book and movie, *The
Exorcist.* Explores this social phenomenon, Bible teach-
ings, demon activity, and the psychological effects. An
intriguing search for truth about supernatural forces.
                                                   No. 036, Paper $1.95

DINOSAURS: Those Terrible Lizards
Duane T. Gish, Ph.D.; Illustrated by Marvin Ross

At last! A book for young people, from a creationist perspective, on those intriguing dinosaurs. No one living in the world today has ever seen a real live dinosaur—but did people in earlier times live with dinosaurs? Were dragons of ancient legends really dinosaurs? Does the Bible speak about dinosaurs? The answers are in this book! Written by Dr. Gish, noted scientist who is author of the best seller *EVOLUTION? The Fossils Say NO!*, this book is profusely illustrated in color on beautiful 9" x 11" pages.   No. 046, Cloth $5.95

EVOLUTION? The Fossils Say NO!
Duane T. Gish, Ph.D.

The fossil record proves there has been no evolution in the past and none in the present, conclusively documented in this critique. Many photos. Over 100,000 in print.

No. 054, Paper $1.95

PUBLIC SCHOOL EDITION (non-religious text)   No. 055, Paper $2.95

Scientific Creationism
Ed. by Henry M. Morris, Ph.D.

The most comprehensive, documented exposition of all the scientific evidences dealing with origins, showing clearly the superiority of the creation model over the evolution model. Intended especially as a reference handbook for teachers.   No. 140, Kivar $4.95

PUBLIC SCHOOL EDITION (non-religious text)

No. 141, Kivar $5.95; No. 357, Cloth $7.95